42L XX40OX

A CONCEPT
OF LIMITS

Donald W. Hight
Department of Mathematics
Pittsburg State University

Dover Publications, Inc., New York

To My Parents—
My Appreciation of Them Is
Monotone Increasing and Unbounded Above

Published in Canada by General Publishing Company, Ltd., 30 Lesmill Road, Don Mills, Toronto, Ontario.
Published in the United Kingdom by Constable and Company, Ltd., 10 Orange Street, London WC2H 7EG.

This Dover edition, first published in 1977, is a corrected republication of the work first published by Prentice-Hall, Inc., Englewood Cliffs, N. J., in 1966. The section entitled "Answers to Even-Numbered Exercises," published in 1966 by Prentice-Hall as a separate supplement, is here republished as an integral part of the book.

International Standard Book Number: 0-486-63543-0
Library of Congress Catalog Card Number: 77-80029

Manufactured in the United States of America
Dover Publications, Inc.
180 Varick Street
New York, N.Y. 10014

Contents

Preface

LIMITS HAVE occupied a unique position in mathematics education. In secondary school or precalculus mathematics, they have been regarded as "too hard" and thus avoided; in calculus courses, they have been taken for granted or skimmed over hurriedly as if they were "intuitively obvious." The purpose of this book is to extend a concept of limits from intuitive ideas about limits to knowledge of a generalized limit that is applicable in many areas of mathematics. To achieve this goal, a method has been employed that provides understanding through participation. Considerable effort has been made to allow and encourage each reader to progress at his own rate through a sequence of developmental steps from a known concept to a desired conclusion. Therefore, the book begins very slowly and assumes knowledge of high school algebra and an acquaintance with trigonometry. Then, it progresses through historical accounts of limits, the limits of sequences and functions, continuity, and proofs of theorems to the development and applications of a unifying concept of a generalized limit.

In addition to an understanding of a concept of limits, the book offers the following: motivated review and reinforcement of familiar topics such as algebra, inequalities, absolute value, functions, graphs, trigonometry, series, and geometry related to the measure of a circle; a thorough mathematical study of an important concept; the appreciation of definitions, proofs, and a mathematical structure; transition from typical precalculus and traditional mathematics to more sophisticated contemporary analysis; and some experience at "limit-guessing" and "δ-finding."

Chapter 1 is devoted to an intuitive and historical development of the limit of a sequence. The definition of a limit is preceded by "Examples to Ponder," geometrical and historical examples, graphs, and repeated questions such as, "For what values of n is s_n within ϵ of L?" Experience with the concept is provided by some simple theorems, proofs that some real number L is or is not the limit of a sequence s, and helpful applications.

Chapter 2 presents standard limits of functions and continuity. Although a sequence was called a function in Chapter 1, functions are presented in greater detail in this section. Stress is placed upon simple theorems concerned with boundedness, uniqueness of limits, and positive or negative limits. Examples and exercises emphasize the interrelation of various types of limits, special limits, and continuity of trigonometric functions.

In Chapter 3, unifying attributes of the defined limits of sequences and functions are organized into a generalized limit. Then, by proving a single theorem for the generalized limit, a proof is given that is readily applicable to every type of limit previously discussed. The chapter also introduces limits and continuity of sums, differences, products, quotients, and composites of functions and concludes with applications of limits to high school mathematics, to calculus, and to extensions of the generalized limit theorems.

Mathematical terms that are especially associated with the presentation are written in boldface when they are introduced; similarly, important terms that have more widespread usage are written in italics.

Convention has been broken when it was thought that better understanding would result. Such expressions as "$\lim_{b} \{(x, f(x))\}$" with "$\lim_{b} f(x)$" and "$\lim_{b} \{(n, s_n)\}$" with "$\lim_{b} s_n$" are used to stress that a limit of a function or sequence has been defined and not a limit of a set of range values. Since students today may not think of a variable as a "quantity that changes," reference to "active" variables such as "x increases without bounds," "$f(x)$ approaches 1," "$x \to \infty$," and "$f(x) \to 1$" are avoided. Instead, sets of numbers and their relationship are used to present the ϵ, δ concept throughout, from intuitive examples to definitions. Furthermore, such names as "limit at b" and "limit at-the-right" are used for "limit as x approaches b" and "limit as x tends toward infinity." Another innovation, graphs of sequences on "n-inverted" coordinate systems, has proved to be quite successful in practice. Also, the presentation of a generalized limit not only allows opportunity to unify the concept, but the proof of one generalized limit theorem provides proofs of theorems for six types of familiar limits plus other limits that readers may subsequently encounter.

I wish to express my gratitude and indebtedness to Dr. Bruce E. Meserve for his constructive criticism, guidance, and encouragement during the preparation of the book. Also, thanks are due to Dr. R. G. Smith, Dr. Glen Haddock, to my wife Betty, and to the many others who assisted me in preparing and testing this material.

Donald W. Hight

chapter 1

Sequences and Their Limits

This book is intended to be read with a pencil in hand. It is not designed to be read as a story, for unless you knew in advance about limits you would soon be confused and lost. Many examples are given which you should analyze and classify. Questions are asked which you are to ponder and answer for yourself. It will then be possible (let us hope) for you to anticipate subsequent considerations and eventually to grasp for yourself a limit concept and to obtain for yourself acceptable definitions of limits. Now, if you have not already done so, get a pencil in hand and a pad of paper beside your book so that we may start our explorations.

1-1 Infinite Sequences

You already have some idea of what a sequence is, for the word is common. In referring to a *sequence* of events you want to communicate that one event happened, then the next, and the next, and so forth. We wish to define an infinite numerical sequence in a similar but more specific manner. To specify that there is a first event and then a next event and a next, and so forth, we utilize the natural numbers (or the set of positive integers). The "events" that we consider are real numbers and are called **terms** of the sequence. An **infinite sequence** of real numbers is a function in which each natural number is associated with a unique real number. Since we are concerned in this text only with infinite sequences of real numbers, we shall refer to them simply as **sequences**.

We shall express a sequence in a traditional manner as an *ordered set* or *list* and also as a set of *ordered pairs*. Thus a sequence s may be expressed either as a list or ordered set,

$$s_1, s_2, s_3, \ldots, s_n, \ldots,$$

or as a set of ordered pairs of related numbers,

$$\{(n, s_n)\}.$$

Here as throughout this book, we shall use the letter "n" as a symbol for a natural number. Thus, each expression for the sequence indicates that 1 is associated with s_1, 2 is associated with s_2, 3 is associated with s_3, and, in general, every natural number n is associated with a unique real number s_n.

Example 1 Express (**a**) as a list and (**b**) as a set of ordered pairs the sequence in which $s_n = 2n$.

> *Solution:* (**a**) 2, 4, 6, 8, 10, ..., $2n$, ... (**b**) $\{(n, 2n)\}$; the set which consists of the ordered pairs (1, 2), (2, 4), (3, 6), and in general $(n, 2n)$ for every natural number n.

Not all sequences can conveniently be expressed by a simple algebraic equation involving n. Some are better suited to a general expression accompanied by an explanation.

Example 2 Express (**a**) as a list and (**b**) as a set of ordered pairs the sequence in which each odd natural number $\mathcal{O}$ is associated with $2(\mathcal{O} + 1)$ and each even natural number e is associated with $2(e - 1)$.

> *Solution:* (**a**) 4, 2, 8, 6, 12, 10, ..., s_n, ..., where $s_n = 2(n + 1)$ if n is odd, and $s_n = 2(n - 1)$ if n is even. (**b**) $\{(n, s_n)\}$ where $s_n = 2(n + 1)$ if n is odd and $s_n = 2(n - 1)$ if n is even.

The **domain** of a sequence (infinite sequence) is the set of natural numbers. The **range** of a sequence is the set of terms of the sequence. We should recall the accepted usage of braces to express a set and be reminded that an expression of a set does not indicate any ordering of its elements. Thus the domain of every sequence is generally expressed as $\{1, 2, 3, 4, \ldots, n, \ldots\}$, where the natural numbers are listed in their natural order by habit and for convenience. We are denoting the set of natural numbers and nothing else. The ranges of the sequences in Examples 1 and 2 are the same set, the set of even natural numbers. This set could be expressed as $\{2, 4, 6, 8, \ldots, 2n, \ldots\}$ or simply as S if we defined S to be the set of even natural numbers.

Two sequences s and t are **equal** if and only if $s_n = t_n$ for every natural number n. The sequences in Examples 1 and 2 are sequences that are not equal. However, their domains are equal (the same set—the set of natural numbers) and their ranges are equal (the same set—the set of even natural numbers).

Example 3 **(a)** List the first five terms of the sequence $\{(n, s_n)\}$ in which $s_n = (-1)^n$. **(b)** Give the range of the sequence.

> *Solution:* **(a)** Note that $(-1)^1 = -1$, $(-1)^2 = 1$, $(-1)^3 = -1$, $(-1)^4 = 1$, and $(-1)^5 = -1$. Hence, the first five terms in order are $-1, 1, -1, 1, -1$. **(b)** The range is the finite set $\{-1, 1\}$ since $(-1)^n = -1$ if n is odd and $(-1)^n = 1$ if n is even.

Exercises

1. List the first five terms of the sequence $s = \{(n, s_n)\}$ in which $s_n = 2n - 1$.

2. List the first five terms of the sequence $t = \{(n, t_n)\}$ in which $t_n = 1/n$.

3. Consider the sequence $c = 2, 2, 2, 2, \ldots, 2, \ldots$ in which $c_n = 2$ for each n. **(a)** State the range of the sequence c. **(b)** Represent the sequence in the form $\{(n, s_n)\}$.

4. Give a representation in the form $\{(n, s_n)\}$ of a sequence whose range is $\{1\}$. (This sequence and the sequence in Exercise 3 are called **constant sequences** because they may be represented in the form $\{(n, c)\}$ for some constant $c = c_n$ for every natural number n.)

5. Consider the sequence $1, 2, 3, 3, \ldots, 3, \ldots$ in which $s_1 = 1$, $s_2 = 2$, and $s_n = 3$ for each natural number $n \geq 3$. Name two other sequences that have the same range.

6. Let the first term of a sequence $\{(n, a_n)\}$ be some given real number, say c, and let $a_n = c + (n - 1)d$, where d is a given real number. Write the first five terms of this sequence. (Such a sequence is called an **arithmetic sequence** and can be expressed by "$\{(n, c + (n - 1)d)\}$.")

7. Let the first term of a sequence g be some given real number, say a where $a \neq 0$, and let $g_n = ar^{n-1}$, where r is a given real number. Write the first five terms of the sequence g. (Such a sequence is called a **geometric sequence** and is expressed by "$\{(n, ar^{n-1})\}$.")

8. Let the first two terms of a sequence s each be 1 and $s_n = s_{n-1} + s_{n-2}$ for each natural number $n \geq 3$. For example, $s_3 = 1 + 1 = 2$. Write the first seven terms of this sequence. (This sequence is called a **Fibonacci sequence**.)

1-2 Graphs of Infinite Sequences

To stimulate your intuition and sharpen your perception let us consider some graphs of sequences. The **graph of a sequence** s is the set of all points whose coordinates are (n, s_n) on a coordinate system with an n-axis and an s_n-axis. As you make graphs and look at them, do not hestitate to look for new "truths" or relationships. However, do not be completely satisfied that the impressions of your eye are valid until a proof is given using previously accepted theorems, definitions, or properties of numbers.

Example 1 Graph on a Cartesian coordinate system the sequence $\left\{\left(n, \dfrac{1}{n}\right)\right\}$.

Solution: We select the coordinate axes on perpendicular real number lines that intersect at the point with coordinates $(0, 0)$. Let the n-axis be the horizontal line and the s_n-axis be the vertical line as in Figure 1-1. The graph of the number pair (n, s_n) will be a point at a

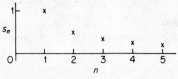

Figure 1-1

distance n units to the right of 0 and at a distance $|s_n|$ up or down from 0 (up if $s_n > 0$, down if $s_n < 0$, 0 if $s_n = 0$).

Example 2 Graph on a Cartesian coordinate system the sequence $\{(n, 1)\}$.

Solution:

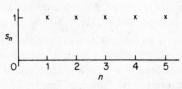

Figure 1-2

The sequences graphed in Examples 1 and 2 are infinite sequences and their graphs are infinite sets of points. The graphs extend indefinitely off the paper because our sequences are infinite sequences, and because we associate with each n the point whose distance is n units to the right of 0. One way that this difficulty can be overcome for infinite sequences is by associating with each natural number n the point whose distance is $1/n$ units to the left of 0. In this scheme the graph of the number pair (n, s_n) will be a point at a distance $1/n$ units to the left of 0 and at a distance $|s_n|$ up or down from 0. Henceforth in this book such a graph will be referred to as a graph on an **n-inverted coordinate system**.

Example 3 Graph on an n-inverted coordinate system the sequence $\left\{\left(n, \dfrac{1}{n}\right)\right\}$.

Solution: We indicate the points on the *n*-axis whose distance to the left of 0 is $1/n$ by "$-1/n$." Also, we indicate the natural numbers that we associate with these points.

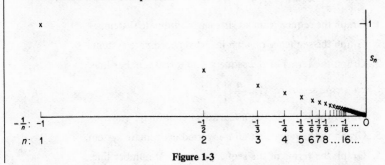

Figure 1-3

Example 4 Graph on an *n*-inverted coordinate system the sequence $\{(n, 1)\}$.

Solution: Since the location on the *n*-axis of a point associated with a natural number *n* should now be understood, we indicate these points with natural numbers only.

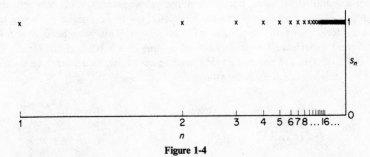

Figure 1-4

Frequently it is instructive to graph a sequence on a real number line. In such a graph only the terms are plotted on the number line, and the relationship between *n* and s_n is not discernible unless each point is labeled. This type of graph is shown in Figure 1–5, but is not used as frequently as the other two types in the remainder of this book.

Example 5 Graph the sequence $\left\{\left(n, \dfrac{1}{n}\right)\right\}$ on a real number line.

Solution:

Figure 1-5

Exercises

Consider the sequence $\left\{\left(n, \dfrac{2}{n}\right)\right\}$:

1. Graph the sequence on a Cartesian coordinate system.
2. Graph the sequence on an n-inverted coordinate system.
3. Graph the terms of the sequence on a real number line.

Consider the sequence $\left\{\left(n, \dfrac{n}{2}\right)\right\}$:

4. Graph the sequence on a Cartesian coordinate system.
5. Graph the sequence on an n-inverted coordinate system.
6. Graph the terms of the sequence on a real number line.

1-3 Some Examples to Ponder

We shall consider three groups of sequences. You should look for similarities and differences between individual sequences and between the groups of sequences; you should, especially, try to determine the common characteristics of the sequences of each group. This may be a new and different experience for you as a reader—but it is a basic characteristic of this book. You are not being told anything about the sequences; you should investigate them carefully by yourself.

	Group I	Group II	Group III
Ia.	$\left\{\left(n, (-1)^n \dfrac{1}{n}\right)\right\}$	IIa. $\{(n, n)\}$	IIIa. $\{(n, (-1)^n)\}$
Ib.	$\left\{\left(n, \dfrac{n-1}{n}\right)\right\}$	IIb. $\{(n, -n)\}$	IIIb. $\left\{\left(n, (-1)^n \dfrac{n-1}{n}\right)\right\}$
Ic.	$\left\{\left(n, \left(-\dfrac{1}{2}\right)^n\right)\right\}$	IIc. $\{(n, n^2)\}$	IIIc. $\{(n, (-1)^n n)\}$
Id.	$\left\{\left(n, 1 + \dfrac{1}{n}\right)\right\}$	IId. $\{(n, n - 10)\}$	IIId. $\{(n, (-2)^n)\}$

If you have had trouble getting started with your analysis of the sequences, try writing down the first few terms of each sequence and graph each sequence in a couple of ways. For example, you might notice that the sequences Ib, IIa, and possibly others, have the property that $s_1 < s_2 < s_3 < \ldots < s_n < \ldots$. Also, Ia, Ib, Ic, and possibly others, have the property that each term is less than or equal to 1. You should make a list for yourself of the similarities and the differences that you have observed before reading further.

Now that you have written a few terms of each sequence and have graphed each one at least in the two ways suggested in Examples 1 and 3 of §1–2, you should feel more familiar with the sequences. Next, try to add new sequences to each group so that each new sequence has the properties of the others in its group. If you have not at least determined some terms, made graphs, and thought about the sequences, you should go back now and do it. Learning mathematics is like eating—there is no satisfaction in watching someone else do it or assuming you could do it if you wanted. In order to gain full benefit from this book you should go back now and analyze or check your analysis of the sequences in Groups I, II, and III. Then keep your analysis and graphs at hand, and complete the following exercises.

Exercises

Each of the following exercises refers to the sequences in Groups I, II, *and* III:

1. Identify the sequences which have terms that are (**a**) greater than 5; (**b**) greater than 25; (**c**) greater than 100; (**d**) greater than any specific real number that can be named.

2. Identify the sequences which have terms that are (**a**) less than -10; (**b**) less than -100; (**c**) less than any specific number that can be named.

3. Identify sequences which have their entire set of terms between 100 and -100.

4. For each of the three sequences given below, name numbers N and M, where $M \le N$, such that each term s_n of the sequence has the relationship $M \le s_n \le N$.

   (**a**) Ia. $\left\{\left(n, (-1)^n \frac{1}{n}\right)\right\}$.   (**b**) Ib. $\left\{\left(n, \frac{n-1}{n}\right)\right\}$.

   (**c**) IIIb. $\left\{\left(n, (-1)^n \frac{n-1}{n}\right)\right\}$.

5. Identify the sequences for which two numbers N and M can be given such that each term s_n will have the relation $M \le s_n \le N$.

6. Identify the sequences for which $s_{n+1} \le s_n$ for every natural number n.

7. Identify the sequences for which $s_{n+1} \ge s_n$ for every natural number n.

1-4 Types of Sequences

If you analyzed the examples of §1–3, you probably noted some properties held in common by some sequences. In this section it is our intention to give commonly accepted names for the properties that you were apt to notice.

For our convenience in making definitions we shall use *iff* to mean *if and only if*. Also, as previously stated, the symbol "n" is to represent a natural number and the other letters represent real numbers unless otherwise specified.

The first set of definitions refers to a property that should be associated with the existence of a "highest" or "lowest" point on the graph of a particular sequence. A sequence is **bounded above** iff there exists a real number N such that $s_n \leq N$ for every n. The number N is called an **upper bound** of s. A sequence s is **bounded below** iff there exists M such that $M \leq s_n$ for every n. The number M is called a **lower bound** of s. A sequence s is **bounded** iff it is bounded above and bounded below.

Example 1 Prove that 2 is an upper bound of $\{(n, (n-1)/n)\}$.

> *Solution:* It must be shown that $\dfrac{n-1}{n} \leq 2$ for every natural number n. Notice first that $\dfrac{n-1}{n} = 1 - \dfrac{1}{n}$ and second that $-\dfrac{1}{n} < 0$ since $n \geq 1$. Therefore, $\dfrac{n-1}{n} = 1 - \dfrac{1}{n} < 1$. Hence, 2 is an upper bound of the sequence, and indeed 1 is also an upper bound that is smaller than 2.

As this example suggests, it is possible for a sequence to have a multitude of upper bounds. Indeed, if 2 is an upper bound of a sequence s, then every real number greater than 2 would also be an upper bound of s. Every real number less than 2, however, could not be an upper bound. Therefore, if a sequence s has an upper bound, we frequently find ourselves seeking a smallest or least upper bound. A number m is a **least upper bound** of a sequence s iff m is an upper bound of s and $m \leq m'$ where m' is any upper bound of s. Although we have not verified that every bounded sequence has a least upper bound, we can prove that it is unique if it exists.

Theorem 1-4a *If a sequence has a least upper bound, then it has only one least upper bound.*

Proof: A sequence s is given with least upper bound m. We assume that m' is a least upper bound of s and prove that $m = m'$. If $m < m'$, then m' would not be a least upper bound; and likewise, if $m' < m$, then m would not be a least upper bound. Hence, $m \geq m'$ and $m' \geq m$, and therefore $m = m'$. Any two least upper bounds of a sequence s are equal; if a sequence has a least upper bound, it has only one.

A similar definition and theorem could be formed for **greatest lower bound**. These are left as exercises (Exercises 7a and 8).

If you made your graphs in § 1–3 by plotting $(1, s_1)$, $(2, s_2)$, $\ldots$, (n, s_n), $(n + 1, s_{n+1})$, $\ldots$ in the order specified by the domain, then the next set of definitions refers to a property that should be associated with the height of each point relative to the previously plotted point. A sequence is **monotone increasing** iff $s_{n+1} > s_n$ for every n. A sequence is **monotone nondecreasing** iff $s_{n+1} \geq s_n$ for every n. In a similar manner **monotone decreasing** and **monotone nonincreasing** could be defined. This, however, is left as an exercise (Exercise 7). A sequence is called **monotone** if it is any one of the four types mentioned above.

In some exercises that follow, the instruction "classify according to boundedness" means to determine whether or not each sequence is bounded above or below. Similarly, the instruction "classify according to monotonicity" means to determine whether or not each sequence is monotone increasing, decreasing, nonincreasing, or nondecreasing.

Example 2 Classify the sequence $\left\{ \left(n, \dfrac{n-1}{n} \right) \right\}$ according to (a) boundness; (b) monotonicity. Also, (c) prove your answer to part (b).

> *Solution:* (a) Bounded (above by 1, below by 0). (b) Monotone increasing and, consequently, monotone nondecreasing. (c) For every natural number n, $(n + 1) > n$;
>
> $$\frac{1}{n} > \frac{1}{n+1}; \qquad 1 - \frac{1}{n} < 1 - \frac{1}{n+1}; \qquad \frac{n-1}{n} < \frac{(n+1)-1}{n+1}.$$
>
> For every natural number n, $s_n < s_{n+1}$ and $s_n \leq s_{n+1}$; the sequence is monotone increasing and monotone nondecreasing.

Exercises

Each of the following exercises refers to the sequences in Groups I, II, *and* III *of* § 1–3.

1. Classify each of the sequences according to monotonicity.

2. Classify each of the sequences according to boundness.

3. (a) Is 100 an upper bound of the sequence Ia? (b) Is 99? (c) Is 98? (d) Are there numbers less than 98 that are upper bounds of the sequence Ia? (e) What number is the least upper bound of the sequence Ia?

4. Find the least upper bound, if there is any, of each of the sequences in § 1–3.

5. (a) Is 0 a lower bound of the sequence Id? (b) Is $\frac{1}{2}$? (c) Is 1? (d) Is there a number greater than 1 that is a lower bound of the sequence Id?

6. Find the greatest lower bound, if there is any, of each of the sequences in § 1–3.

7. Give definitions of the following terms for a sequence s: (a) greatest lower bound; (b) monotone decreasing; (c) monotone nonincreasing.

8. Prove that if a sequence s has a greatest lower bound m, then it has only one greatest lower bound.

1-5 Circumference of a Circle

The problem of determining the circumference of a circle is different from finding the perimeter of a polygon. The perimeter of a polygon can be found by determining the length of each side and then adding these lengths. A circle is not a polygon; it does not have line segments as "sides." However, it is possible to relate perimeters of certain polygons to the circumference of a circle. Archimedes (287–212 B.C.) used sequences to study the circumferences of circles and to determine the number π. We now generate a sequence that may be used for this purpose much as Archimedes did over 2000 years ago.

For purposes of demonstration consider a circle with a diameter of one unit in length. Suppose first that a square is inscribed in the circle as in Figure 1–6. Because the shortest distance between two points is measured along a straight line, the perimeter of the inscribed square is less than the circumference of the circle. An inscribed regular octagon which shares four vertices of the square has a perimeter larger than that of the square and less than the circumference of the circle (Figure 1–6). Consider a sequence of perimeters of inscribed regular polygons of 4, 8, 16, 32, . . . , 2^{n+1}, . . . sides in which each polygon except the square shares the vertices of the preceding polygon.

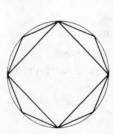

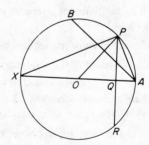

Figure 1-6   Figure 1-7

To determine the perimeters of the polygons we must utilize some concepts of plane geometry. Therefore, as in Figure 1–7, let AOX be the diameter of a circle with radius r. Let $\overline{AB}$ be a side of a regular polygon of 2^{n+1} sides, each of length s_n, and $\overline{AP}$ be a side of a regular polygon of $2^{(n+1)+1}$ sides, each of length s_{n+1}.

Construct chord $\overline{PR} \perp \overline{AX}$ and let $\overline{PR}$ intersect $\overline{AX}$ at Q. Draw $\overline{AR}$.

Then, $\overline{AP} = \overline{AR}$ since the diameter $\overline{AX}$ bisects the arc $\overset{\frown}{PR}$, and the equal arcs $\overset{\frown}{AP}$ and $\overset{\frown}{AR}$ have equal chords. Also, $\overline{AB} = \overline{PR}$ since $\overset{\frown}{AB} = 2\overset{\frown}{AP} = \overset{\frown}{PR}$. Draw $\overline{PO}$ and $\overline{PX}$. Now $\sphericalangle APX$ is inscribed in a semicircle and is a right angle. Right triangle APX has altitude $\overline{PQ}$ and, therefore, $\overline{AP}$ is the mean proportional between $\overline{AQ}$ and $\overline{AX}$. If we let the symbol "$\overline{AP}$" stand for the line segment AP and also for the length of $\overline{AP}$ and do likewise with $\overline{AX}, \overline{AQ}$, and so forth,† then the proportion can be written in the form of an equation:

$$\overline{AP}^2 = \overline{AX} \cdot \overline{AQ}. \tag{1}$$

Since triangle QOP is a right triangle, the Pythagorean relation yields

$$\overline{AO} - \overline{AQ} = \overline{QO} = \sqrt{\overline{OP}^2 - \overline{QP}^2}. \tag{2}$$

If we solve equations (1) and (2) simultaneously to eliminate $\overline{AQ}$, we get

$$\overline{AP}^2 = \overline{AX}(\overline{AO} - \sqrt{\overline{OP}^2 - \overline{QP}^2})$$

or

$$\overline{AP}^2 = \overline{AO} \cdot \overline{AX} - \overline{AX}\sqrt{\overline{OP}^2 - \overline{QP}^2}.$$

If we note that $\overline{AP} = s_{n+1}$, $\overline{AO} = \overline{OP} = r$, $\overline{AX} = 2r$ and $\overline{QP} = \frac{1}{2}s_n$, then $\overline{AX}\sqrt{\overline{OP}^2 - \overline{QP}^2} = 2r\sqrt{r^2 - (\frac{1}{2}s_n)^2} = r\sqrt{4r^2 - (s_n)^2}$, $\overline{AO} \cdot \overline{AX} = 2r^2$, and we obtain

$$s_{n+1} = \sqrt{2r^2 - r\sqrt{4r^2 - (s_n)^2}}. \tag{3}$$

The perimeter of a square is $2\sqrt{2}$ units if it is inscribed in a circle of diameter 1 unit (that is, $r = \frac{1}{2}$ unit). The length of a side of an inscribed regular octagon is determined by (3) from the square. Repeated use of (3) will generate a sequence of lengths of sides and hence a sequence of perimeters as indicated by the chart below.

n:	1,	2,	3,	4,	...
No. of sides, 2^{n+1}:	4,	8,	16,	32,	...
Length of side, s_n:	$\frac{1}{2}\sqrt{2}$,	$\frac{1}{2}\sqrt{2-\sqrt{2}}$,	$\frac{1}{2}\sqrt{2-\sqrt{2+\sqrt{2}}}$,	$\frac{1}{2}\sqrt{2-\sqrt{2+\sqrt{2+\sqrt{2}}}}$,	...
Perimeter, (approximate) p_n:	2.8284,	3.0616,	3.1208,	3.1360,	...

Figure 1-8

Although the successive terms of the sequence indicated in the chart are assured, the actual calculation of the decimal approximation becomes increasingly difficult. The purpose of the example was not to calculate terms but rather to generate a very special pair of sequences $\{(n, s_n)\}$ and $\{(n, p_n)\}$.

---

† Since there is no one standard symbol for segment AP or for the length of segment AP, this procedure is appropriate for our uses.

Exercises are provided to encourage you to reconsider the process outlined above and to assist you in comprehending and appreciating the use of the generated sequences.

Exercises

Exercises 1 through 4 refer to the sequences $\{(n, s_n)\}$ *and* $\{(n, p_n)\}$ *in Figure 1–8.*

1. Classify the sequence $\{(n, s_n)\}$ according to (a) boundedness; (b) monotonicity.

2. Classify the sequence $\{(n, p_n)\}$ according to (a) boundedness; (b) monotonicity.

3. If the diameter of a circle is 1 unit, then the circumference is πd units $= \pi$ units. Will any term of $\{(n, p_n)\}$ be greater than or equal to π?

4. What is the relationship of the sequence $\{(n, p_n)\}$ to the real number π when the circle has a diameter of 1 unit?

5. If a sequence $\{(n, p'_n)\}$ of perimeters of circumscribed regular polygons of 4, 8, 16, ..., 2^{n+1}, ... sides were formed, how would it be classified according to (a) boundedness; (b) monotonicity?

6. Determine the greatest lower bound of the sequence $\{(n, p'_n)\}$.

7. If a sequence $\{(n, A_n)\}$ of areas of the inscribed regular polygons of 4, 8, 16, ..., 2^{n+1}, ... sides were formed, how would it be classified according to (a) boundedness; (b) monotonicity?

1-6 Sequences in the Physical World and in History

Zeno of Elea (495–435 B.C.) stimulated thinking on the part of men from his day to this with his famous paradoxes. Of the eight that are recorded in the writing of Aristotle, the most famous one concerns sequences and involves a race between Achilles and a tortoise.

For an example of the paradox, think of the race with reference to a number line on which Achilles starts at the origin and the tortoise starts at the point with coordinate 1. Suppose for convenience that Achilles runs twice as fast as the tortoise crawls (a slow track man or a speedy turtle). Let the rate of the tortoise be 1 unit per second and that of Achilles be 2 units per second.

When time is 0 the race starts with Achilles at the origin and the tortoise at the point with coordinate 1. Achilles will need $\frac{1}{2}$ second to cover 1 unit of distance at a rate of 2 units per second. During this time the tortoise has traveled only $\frac{1}{2}$ unit. Hence, after $\frac{1}{2}$ second Achilles is only $\frac{1}{2}$ unit behind the tortoise who is at the point with coordinate $1\frac{1}{2}$. In the next $\frac{1}{4}$ second Achilles moves $\frac{1}{2}$ unit and the tortoise moves $\frac{1}{4}$ unit. Proceeding in this manner, we generate three sequences as shown in Figure 1–9.

t (time in seconds):	$0, \ \frac{1}{2}, \ \frac{3}{4}, \ \frac{7}{8}, \ \frac{15}{16}, \ \ldots, t_n, \ldots$	where $t_n = 1 - (\frac{1}{2})^{n-1}$
a (Achilles' position):	$0, \ 1, \ 1\frac{1}{2}, \ 1\frac{3}{4}, \ 1\frac{7}{8}, \ \ldots, a_n, \ldots$	where $a_n = 2 - (\frac{1}{2})^{n-2}$
r (tortoise's position):	$1, \ 1\frac{1}{2}, \ 1\frac{3}{4}, \ 1\frac{7}{8}, \ 1\frac{15}{16}, \ \ldots, r_n, \ldots$	where $r_n = 2 - (\frac{1}{2})^{n-1}$

Figure 1-9

When Achilles reaches 1, the tortoise is at $1\frac{1}{2}$. When Achilles reaches $1\frac{1}{2}$, the tortoise is at $1\frac{3}{4}$. Achilles must reach the spot where the tortoise just was, but during this time the tortoise has moved forward. Hence, Zeno concluded, Achilles can never catch the turtle. What do you think?

For another familiar example, think of the action of a dropped rubber ball. We shall assume that a ball (theoretically, at least) is dropped from a distance of 1 unit to a flat surface and that it rebounds to exactly one-half its previous height. The action of such a ball, suggested in Figure 1–10, enables us to generate a sequence h of successive heights reached by the ball. Furthermore, we can determine a sequence b of numbers b_n representing the distance the ball has traveled since being dropped until it hits the ground for the nth time,

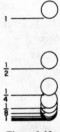

Figure 1-10

$$b_n = h_1 + 2h_2 + 2h_3 + \ldots + 2h_n.$$

The two sequences h and b are exhibited in Figure 1–11.

n:	$1, \ 2, \ 3, \ 4, \ \ldots,$	$n, \ \ldots$
h (height):	$1, \ \frac{1}{2}, \ \frac{1}{4}, \ \frac{1}{8}, \ \ldots,$	$(\frac{1}{2})^{n-1}, \ldots$
b (distance):	$1, \ 2, \ 2\frac{1}{2}, \ 2\frac{3}{4}, \ \ldots,$	$3 - (\frac{1}{2})^{n-2}, \ldots$

Figure 1-11

How would you answer the question: "How far does the bouncing ball travel?"

The principle of the pendulum was supposedly discovered by Galileo Galilei in 1583 as he watched a light swinging on a chain in the cathedral in Pisa. He proposed that the time it takes for a pendulum to swing through one cycle is independent of the amplitude of the swing. To demonstrate his principle and to generate a sequence for study, we shall construct a hypothetical situation involving a pendulum and apply Galileo's principle.

Let the period of a pendulum (the time it takes it to swing over and back) be 4 seconds. Suppose it slows down in such a way that the displacement at each apex is $\frac{9}{10}$ of the displacement of the preceding apex. We shall use n to represent time in seconds and d_n to represent the displacement of the pendulum at time n (note that $d_n > 0$ if the pendulum is to the right and $d_n < 0$ if it is to the left) as shown in Figure 1–12.

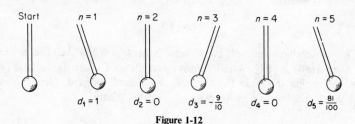

Figure 1-12

The general term of the sequence d can be defined by the compound sentence:

$$d_n = \left(-\frac{9}{10}\right)^{(1/2)(n-1)} \quad \text{if } n \text{ is odd,}$$

and

$$d_n = 0 \quad \text{if } n \text{ is even.}$$

The following exercises refer to the sequences that relate to Zeno's paradox, the bouncing ball, and the pendulum. As statements are made concerning the sequences, think also of the physical actions that are described by the sequences.

Exercises

Let t, a, and r be the sequences in Figure 1–9, let h and b be the sequences in Figure 1–11, and let d be the sequence in Figure 1–12. The following exercises refer to these sequences.

1. Classify each of the sequences according to boundedness.

2. Classify each of the sequences according to monotonicity.

3. Identify the least upper bound of each of the sequences.

4. (a) Does $a_n = r_n$ for some n? (b) Will Achilles catch the tortoise at any time less than 1 second? (c) Where will Achilles be when the time is 1 second? (d) Where will the tortoise be when the time is 1 second? (e) Will Achilles overtake the tortoise?

5. (a) What is the greatest lower bound of the sequence h? (b) If the theoretical ball in Figure 1–10 were at rest relative to the flat surface, where would it be?

6. (a) Is 0 the least upper bound or greatest lower bound of the sequence d?
   (b) If a pendulum were at rest position (with no external influences), what would its displacement be?

1-7 The Limit of a Sequence: Informal

Some examples that have been presented are particularly valuable to relate familiar physical phenomena to an accepted mathematical concept of a limit. The sequence t of units of time in the example of Zeno's paradox has a special relation to the number 1. We call 1 the "limit of the sequence t." The sequences a and r of positions of Achilles and the tortoise both have 2 as a limit.

When we think of the action of a bouncing ball and the sequence h of heights of the ball, we think of 0 as the limit of the sequence h. Similarly, the sequence b of distances that the ball traveled has 3 as its limit.

If a pendulum does not have a sustaining force, the swing gradually dies down. Even though our sequence d of displacements was generated from a theoretical pendulum, do you feel that d has a limit? Note that 0 is neither an upper bound nor lower bound of the sequence d but we would call 0 the limit of the sequence d. A sequence may "reach" or " take on" its limit.

The circumference of a circle is not the perimeter of any regular polygon inscribed in the circle. However, what relationship would you expect between the circumference of a circle and a sequence of perimeters such as $\{(n, p_n)\}$ of § 1–5? It is common to find circumference *defined* as the limit of such a sequence.

Using the intuitive notion of a limit that we formed by the analogies to physical phenomena and geometrical concepts, we should be able to propose limits for various other sequences. The sequences $\{(n, 1)\}$ and $\{(n, 1/n)\}$ of §1–2 and the sequences in Group I of §1–3 each have a limit. You, the reader, should review these sequences and, especially, compare their graphs at this time.

The following example and exercises are designed to permit the discovery of the basic relationship between a sequence and its limit. If you know a definition of a limit, consider it as you work; if you do not know a definition of a limit, seek definitive relationships that will enable you to anticipate a definition.

Example Consider a positive number, say 1/50, and the sequence $\{(n, 1/n)\}$. For what values of n is $1/n$ within 1/100 of 1/50?

   Solution: Note that $(1/50) - (1/100) = 1/100$ and $(1/50) + (1/100)$ $= 3/100$. Hence, $1/n$ is within 1/100 of 1/50 iff

$$\frac{1}{n} > \frac{1}{100} \text{ and } \frac{1}{n} < \frac{3}{100}; \text{ that is, iff } 100 > n \text{ and } \frac{100}{3} < n.$$

Thus, $1/n$ is within $1/100$ of $1/50$ iff $33 < n < 100$.

Exercises

For Exercises 1 through 4 refer to the sequence $\{(n, 1/n)\} = 1, \frac{1}{2}, \frac{1}{3}, \ldots,$ $1/n, \ldots$ *(see Example 1 of* §1–2).

1. For what values of n is $1/n$ within $1/100$ of 0?

2. Let k be any positive integer. For what value of n is $1/n$ within $1/k$ of 0?

3. For what values of n is $1/n$ within $1/100$ of $-1/10$?

4. Compare the answers of the Example, Exercise 1, and Exercise 3.

For Exercises 5 through 8 refer to the sequence $\{(n, (n-1)/n)\} = 0, \frac{1}{2}, \frac{2}{3}, \frac{3}{4},$ $\ldots, (n-1)/n, \ldots$ *(see Ib. of* § 1–3).

5. Let k be any positive integer. For what values of n is $(n-1)/n$ within $1/k$ of 1?

6. For what values of n is $(n-1)/n$ within $1/100$ of 0.9?

7. For what values of n is $(n-1)/n$ within $1/100$ of 1.1?

8. Compare the answers of Exercises 5, 6, 7.

1-8 A Precise Language

To give a definition of the limit of a sequence, we need to have specific, exact meanings for the expressions we use. It is permissible, even desirable, to use numbers and numerical relationships to meet these requirements. Hence, the following symbols and expressions are suggested:

(i) ϵ *(epsilon)—the "closeness" number:* We have been required to find certain terms of a sequence that are within $\frac{1}{10}$ or $\frac{1}{100}$ of some real number. We might have been required to find terms within $\frac{1}{1000}$, $\frac{1}{10^4}$, or $\frac{3}{99997}$ of the real number. In general, we could be required to find terms that are within ϵ of a given real number where ϵ is any positive real number. Hence, we shall use the Greek letter ϵ, called "**epsilon**," as a variable for positive real numbers that generally are required to be less than 1.

(ii) *Absolute value—the distance expression:* Recall that the absolute value of a real number b is positive unless $b = 0$. For example, $|5| = 5$, $|-3| = 3$, $|0| = 0$. A very satisfactory definition of the absolute value of a real number is the following:

$$|b| = b \qquad \text{if} \quad b \geq 0,$$
and
$$|b| = -b \qquad \text{if} \quad b < 0.$$

Note that if $b < 0$, then b is negative and $-b$ (the opposite of b) is then positive.

The **closeness** of two numbers s_n and L, or the **distance** between them, is either $L - s_n$ or $s_n - L$ whichever is not negative. To avoid the detail of subtracting the smaller number from the larger, it is convenient to use $|s_n - L|$ to represent the distance or difference between s_n and L. (When *directed distances* are desired, it is customary to ask for the distance *from s_n to L* which is $L - s_n$, or for the distance *from L to s_n* which is $s_n - L$.)

In (i) the variable ϵ was identified as the symbol for the required closeness of two numbers. In (ii) the absolute value $|s_n - L|$ was suggested for the distance between s_n and L. Hence, to require that s_n and L be within ϵ of each other is to specifically and precisely require that

$$|s_n - L| < \epsilon.$$

Example Consider the sequence $\left\{ \left(n, \dfrac{n-1}{n} \right) \right\}$. For what values of n is $\dfrac{n-1}{n}$ within ϵ of 1?

Solution: In terms of our new notation we need only to solve the inequality $\left| \dfrac{n-1}{n} - 1 \right| < \epsilon$. Note that

$$\left| \frac{n-1}{n} - 1 \right| = \left| 1 - \frac{1}{n} - 1 \right| = \left| -\frac{1}{n} \right| = \frac{1}{n}.$$

Thus $\left| \dfrac{n-1}{n} - 1 \right| < \epsilon$ iff $\dfrac{1}{n} < \epsilon$; that is, iff $n > \dfrac{1}{\epsilon}$.

All that we have done in (i) and (ii) is adopt mathematical methods to make possible a precise statement. A variable ϵ has been introduced and an absolute value has been used. These methods might seem difficult on first contact because they involve new notation. However, you should become familiar with them because they express precisely the same concepts that were considered earlier.

Exercises

1. The question, "For what values of n is $\dfrac{1}{n}$ within ϵ of 0?" can be reworded with an absolute value and an inequality. Do so.

2. For what values of n is $\left| \dfrac{1}{n} - 0 \right| < \epsilon$? (Actually find the set of natural numbers n that solves the inequality.)

3. Consider $\left\{\left(n, \dfrac{2}{n}\right)\right\}$. For what values of n is $\left|\dfrac{2}{n} - 0\right| < \epsilon$?

4. For what values of n is $\left|1 + \dfrac{1}{n} - 1\right| < \epsilon$?

5. Notice that in the Example and in Exercises 2 and 4 the expressions of the form $|s_n - L| < \epsilon$ are valid for $n > 1/\epsilon$. Are all such expressions valid when $n > 1/\epsilon$?

6. Some properties of absolute value might be convenient to use. Below are listed properties that can be proved or referred to at a later stage.

   (a) $|a| = |-a|$ and $|s_n - L| = |L - s_n|$
   (b) $|ab| = |a|\,|b|$
   (c) $|a + b| \leq |a| + |b|$
   (d) $|s_n - L| < \epsilon$ if and only if $L - \epsilon < s_n < L + \epsilon$.
   (e) $||x| - |y|| \leq |x - y|$

1-9 Limit of a Sequence

You have become acquainted with a number of sequences. By looking at their graphs and considering their terms you have anticipated that some of the sequences have limits and some do not. You have become acquainted with the use of the variable ϵ and absolute values so you can form precise algebraic statements. Now you should reconsider some of the sequences that you thought had limits, recall the relationship you discovered between the sequences and their limits, use precise language to express this relationship, and form for yourself a definition of the limit of a sequence.

The following questions and answers are intended to suggest a typical thought process leading to a definition of the limit of a sequence.

   (i) What does "L is the limit of $\{(n, s_n)\}$" mean?
       Answer: s_n approaches L.
  (ii) What do you mean by "approaches"?
       Answer: s_n gets closer and closer to L.
 (iii) What do you mean by "close"?
       Answer: $|s_n - L|$; that is, the distance between s_n and L.
  (iv) How close should they be?
       Answer: Within ϵ.
   (v) How do you express this "closeness"?
       Answer: $|s_n - L| < \epsilon$ where ϵ is any given positive real number.
  (vi) For what terms s_n is $|s_n - L| < \epsilon$?
       Answer: For all terms s_n providing n is sufficiently large.
 (vii) What do you mean by "sufficiently large"?
       Answer: For all n such that $n > d$ where d is some positive real number.
(viii) Upon what does the value of d depend?

Answer: The value of d will depend upon the sequence s and the value of ϵ.

(ix) Can you summarize your ideas now to give a precise definition?

Answer: Yes, L is the limit of the sequence $\{(n,\ s_n)\}$ iff for any given positive real number ϵ ("the closeness number") there is a positive real number d ("the sufficiently large number") such that

$$|s_n - L| < \epsilon \text{ (that is, } s_n \text{ is within } \epsilon \text{ of } L)$$

if $n > d$ (that is, if n is sufficiently large).

Your definition may be longer or shorter than the one we have developed, but be sure that it is equivalent and that you understand our development. For our convenience in this book we shall identify a new variable to be used in the definition of a limit. In the previous discussion we used the letter d to express the requirement that n be "sufficiently large." Henceforth, this requirement will be expressed in terms of the Greek letter δ, called **"delta,"** which will stand for a positive real number. Thus our definition becomes: The **limit of a sequence** $\{(n, s_n)\}$ is the real number L iff for every $\epsilon > 0$ there exists $\delta > 0$ such that

$$|s_n - L| < \epsilon \qquad \text{if } n > \delta.$$

If for a given sequence there is a number L for which the definition holds, we shall write

$$lim\ \{(n,\ s_n)\} = L \text{ or } lim\ s_n = L.$$

In other books you may also see $s_n \longrightarrow L$, or $\lim_{n \to \infty} s_n = L$.

In the limit definition, we have the phrase "for every $\epsilon > 0$." The use of the word "every" in this definition is the same as its use in the song lyric, "For every man there's a woman." To prove that a sequence has a limit we consider an arbitrary (general) $\epsilon > 0$ and seek some corresponding value, that we call δ, for which the conditions set forth in the definition are true. In the following example, we prove that we can associate with each $\epsilon > 0$ the real number $\delta = 1/\epsilon$ and the conditions of the definition will be true. The problem of finding δ when an arbitrary $\epsilon > 0$ is given may range from easy to extremely difficult, depending on the sequence in question and our ingenuity.

Example   Use the limit definition to prove that $lim\ (-\tfrac{1}{2})^n = 0$.

Solution: Let any $\epsilon > 0$ be given. We must determine some positive real number δ such that $|(-\tfrac{1}{2})^n - 0| < \epsilon$ if $n > \delta$. To determine δ, we analyze the inequality we seek to establish: $|(-\tfrac{1}{2})^n - 0| < \epsilon$. Using properties of absolute value and Exercise 6 of this section we establish

$$\left|\left(-\frac{1}{2}\right)^n - 0\right| = \left(\frac{1}{2}\right)^n < \frac{1}{n}.$$

Thus, $|(-\frac{1}{2})^n - 0| < \epsilon$ if $1/n < \epsilon$; that is, if $n > 1/\epsilon$. EUREKA! We shall choose $\delta = 1/\epsilon$ and we can conclude: $|(-\frac{1}{2})^n - 0| < \epsilon$ if $n > \delta$ where $\delta = 1/\epsilon$. Therefore, for every $\epsilon > 0$ there exists $\delta > 0$ ($\delta = 1/\epsilon$) such that $|(-\frac{1}{2})^n - 0| < \epsilon$ if $n > \delta$; by definition, $lim \ (-\frac{1}{2})^n = 0$.

Exercises

1. Use the definition to prove that $lim \ 1/n = 0$. *Hint:* Let an arbitrary $\epsilon > 0$ be given and find a real number δ such that $|1/n - 0| < \epsilon$ if $n > \delta$. You can begin by analyzing what you want: $|1/n - 0| < \epsilon$.

2. Use the definition to show that $lim \ \dfrac{n-1}{n} = 1$. *Hint:* Let any $\epsilon > 0$ be given and find a real number to call δ so that $\left|\dfrac{n-1}{n} - 1\right| < \epsilon$ if $n > \delta$.

3. Use the definition to show that $lim \ 2/n = 0$.

4. Use the definition to show that $lim \ 1/n \neq 1/100$. *Hint:* Find some $\epsilon > 0$ for which there is no δ.

5. Consider $lim \ 1/n = 0$. **(a)** If one person shows that when $n > \delta = 1/\epsilon$ then $|1/n - 0| < \epsilon$, then could another person have a different number for δ such that if $n > \delta$ then $|1/n - 0| < \epsilon$? **(b)** Would $\delta = 2/\epsilon$ be such a number? **(c)** Would $\delta = 1/2\epsilon$ be such a number? **(d)** Would any $\delta > 1/\epsilon$ be such a number?

6. Prove that $1/2^n < 1/n$ for every natural number. *Hint:* Note that $1/2^n < 1/n$ iff $2^n > n$. Therefore, prove that $2^n > n$ for every natural number n by mathematical induction; that is, note that $2^1 > 1$ and prove that $2^{k+1} > k + 1$ whenever $2^k > k$.

1-10 Graphical Interpretation of the Limit

Consider the graph of the sequence $\{(n, 1/n)\}$ and the limit 0. In Figure 1–13 a graph on an n-inverted coordinate system is given with a pair of horizontal broken lines at $\frac{1}{5}$ and $-\frac{1}{5}$.

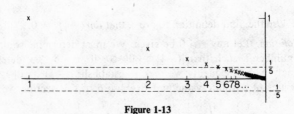

Figure 1-13

Notice that the point whose coordinates are $(5, s_5)$ is plotted on the top horizontal line. Also, every point whose coordinates are (n, s_n) where $n > 5$ is plotted between the horizontal lines. In general, if $\frac{1}{10}$, $\frac{1}{100}$, or any $\epsilon > 0$ is given, a pair of horizontal lines could be drawn at ϵ and $-\epsilon$ about the limit 0. The set of points included between these horizontal lines is called an ϵ-**band** about 0. For every n for which $|s_n - 0| < \epsilon$ the point whose coordinates are (n, s_n) will be plotted in the ϵ-band about 0.

Example 1 Consider the graph in Figure 1–13 of the sequence $\{(n, 1/n)\}$. Discuss the set of points of the graph that are in an ϵ-band about 0 when $\epsilon = \frac{1}{100}$

Solution: Because *lim* $1/n = 0$, all but a finite collection of points of the graph will be in the ϵ-band when $\epsilon = \frac{1}{100}$; that is, since $|s_n - 0| < \frac{1}{100}$ if $n > 100$, all terms except the first 100 will determine points which will be plotted in the ϵ-band when $\epsilon = \frac{1}{100}$.

The graph of a sequence can be very helpful in identifying the limit or some other property of a sequence. Even though looking at or describing a graph may not be a proof of a conjecture, the graph frequently enables us to "try out" a notion and also might suggest a proof to us.

Example 2 Use the graph of the sequence $\{(n, (-1)^n)\}$ to argue that this sequence does not have a limit.

Solution: First, let us form the graph, this time for variety, on a Cartesian system as shown in Figure 1–14. Consider an ϵ-band where $\epsilon = \frac{1}{2}$ whose width is $2 \times \frac{1}{2} = 1$. Is there a fixed real number L and a value for δ such that the points whose coordinates are $\big(n, (-1)^n\big)$ will all be in the ϵ-band about L for every $n > \delta$? Certainly not; no matter what real number is given for δ there will be an *even* integer greater than δ such that $(-1)^n = 1$, and an *odd* integer greater than δ such that $(-1)^n = -1$. Since $|1 - (-1)| > 1$, the points associated with the terms 1 and -1 cannot be plotted in an ϵ-band of any fixed real number L when $\epsilon \leq \frac{1}{2}$. Therefore, the sequence has no limit.

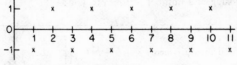

Figure 1-14

Exercises

1. Consider the sequence $\{(n, 1/n)\}$ and its graph in Figure 1–13. Discuss the set of points of the graph that are in an ϵ-band about $\frac{1}{5}$ when

$\epsilon = \frac{1}{10}$; that is, the points between the horizontal lines at $\frac{1}{5} + \frac{1}{10} = \frac{3}{10}$ and $\frac{1}{5} - \frac{1}{10} = \frac{1}{10}$.

2. Consider the sequence $\left\{\left(n, (-1)^n \dfrac{n-1}{n}\right)\right\}$. Is there a fixed real number L, and a value of δ such that $\left|(-1)^n \dfrac{n-1}{n} - L\right| < \epsilon$ for every $n > \delta$ when $\epsilon = \frac{1}{2}$? *Hint:* This sequence is IIIb. of § 1–3 which you have graphed.

A real number c is a cluster point of a sequence s iff for every $\epsilon > 0$ and every natural number N there is some natural number $n > N$ such that $|s_n - c| < \epsilon$.

3. Identify the cluster points of each of the following sequences. The first part is done to exhibit reasoning that can be used to determine cluster points.
   (a) $\{(n, (-1)^n)\}$.
        For any value of N, if $n = 2N$ then $n > N$ and $|s_n - 1| = 0 < \epsilon$; and if $n = 2N + 1$ then $n > N$ and $|s_n - (-1)| = 0 < \epsilon$ for every $\epsilon > 0$. Hence, 1 and -1 are cluster points.
   (b) $\left\{\left(n, (-1)^n \dfrac{n-1}{n}\right)\right\}$;
   (c) Each sequence of Group I of § 1–3.

4. Form (a) a sequence that has 2 and -2 as cluster points; (b) a sequence that has three cluster points.

5. If a sequence has a limit, how many cluster points does the sequence have?

6. If a sequence has two or more cluster points, does the sequence have a limit?

1-11 Theorems on Limits

Can a sequence have two different limits L and M? Suppose $L - M = \frac{1}{10}$, can terms be found to lie within $\frac{1}{100}$ of M and simultaneously lie within $\frac{1}{100}$ of L? If $L - M$ is positive and $\epsilon = \frac{1}{2}(L - M)$ the following graph should suggest a number of ideas to you:

An ϵ-band about L where $\epsilon = \frac{1}{2}(L - M)$	$\left\{\begin{array}{l} L + \frac{1}{2}(L - M) \\ L \\ \frac{1}{2}(L + M) \end{array}\right.$
An ϵ-band about M where $\epsilon = \frac{1}{2}(L - M)$	$\left\{\begin{array}{l} M \\ M - \frac{1}{2}(L - M) \end{array}\right.$

Figure 1-15

If you intend to formulate and prove a theorem for yourself about the uniqueness or singularity of the limit of a sequence, then you should do it before you read on because such a theorem and proof follow:

Theorem 1-11a *If lim $s_n = L$ and lim $s_n = M$, then $L = M$.*

Proof: We shall use the indirect method of proof. Assume $L \neq M$; then one limit is larger than the other. Consider $L > M$ and $L - M > 0$. Since *lim $s_n = L$*, the definition implies that for every $\epsilon > 0$, and in particular for $\epsilon = \frac{1}{2}(L - M)$, there exists $\delta > 0$ such that if $n > \delta$, then $|s_n - L| < \frac{1}{2}(L - M)$ and

$$L - \tfrac{1}{2}(L - M) < s_n < L + \tfrac{1}{2}(L - M). \tag{1}$$

Since *lim $s_n = M$*, then again for every $\epsilon > 0$, specifically $\epsilon = \frac{1}{2}(L - M)$, there exists $\delta' > 0$ such that if $n > \delta'$, then $|s_n - M| < \frac{1}{2}(L - M)$ and

$$M - \tfrac{1}{2}(L - M) < s_n < M + \tfrac{1}{2}(L - M). \tag{2}$$

There are values of n such that $n > \delta$ and $n > \delta'$ for which both (1) and (2) will hold. Hence,

by (1) $\quad \frac{1}{2}(L + M) = L - \frac{1}{2}(L - M) < s_n,$

and by (2) $\quad s_n < M + \frac{1}{2}(L - M) = \frac{1}{2}(L + M).$

Therefore, for some n, $\frac{1}{2}(L + M) < s_n < \frac{1}{2}(L + M)$. But this is impossible because $\frac{1}{2}(L + M) = \frac{1}{2}(L + M)$. Therefore, the assumption $L \neq M$ is false; that is, $L = M$.

In the examples that you studied you probably noted that each sequence that had a limit was a bounded sequence. Will this be the case for every sequence that has a limit? The answer to this question is contained in the statement and proof of the next theorem. You will no doubt note as you read this proof that it depends, basically, on the definition of limit of a sequence and the property that every finite set of real numbers has a largest real number in the set and a smallest real number in the set.

Theorem 1-11b *If a sequence $\{(n, s_n)\}$ has a limit L, then it is a bounded sequence.*

Proof: Select a value for ϵ such as $\frac{1}{10}$. Because the sequence has a limit, we know there is a positive real number δ such that

$$|s_n - L| < \tfrac{1}{10} \quad \text{if} \quad n > \delta;$$

that is, $\quad L - \frac{1}{10} < s_n < L + \frac{1}{10} \quad \text{if} \quad n > \delta.$

Therefore, $\{(n, s_n)\}$ is bounded for all $n > \delta$. To bound the whole sequence we need only to consider $1 \leq n \leq \delta$. However, this is a finite set, and we can select the largest member N and the smallest member M. The sequence $\{(n, s_n)\}$ would then be bounded by the minimum of M and $L - \frac{1}{10}$ and the maximum of N and $L + \frac{1}{10}$.

Exercises

1. Prove that if $lim\ s_n = L > 0$, then there exists a positive real number δ such that if $n > \delta$, then $s_n > 0$.

2. Consider a constant sequence $\{(n, c_n)\}$ where $c_n = c$ for every n. Prove that this sequence has a limit.

3. Given that $lim\ s_n = L$. Consider $\{(n, u_n)\} = s_5, s_6, \ldots, s_{n+4}, \ldots$ where s_1 through s_4 have been omitted. Prove that $\{(n, u_n)\}$ has a limit.

4. Generalize Exercise 3.

5. Given that $lim\ s_n = L$. Consider $\{(n, t_n)\} = x_1, x_2, x_3, s_1, s_2, \ldots, s_{n-3}, \ldots$ where the x's are real numbers. Prove that $\{(n, t_n)\}$ has a limit.

6. Generalize Exercise 5.

7. Given that $lim\ s_n = L$. If $L < b_n < s_n$ for every n, prove that $lim\ b_n = L$. (This statement is frequently treated as a theorem and is an application of the *domination principle* for sequences.)

1-12 Applications of Limits of Sequences

The limit concept is essential to any person, especially to a student of college preparatory mathematics, who seeks an understanding of the real number system and its distinguishing characteristics. Many pre-calculus algebra and geometry textbooks of recent years have pointed out the similarities and differences of the rational number system and the real number system, and some involve the student in a development of the rational number system but not in the development of the real number system.

What are the real numbers? One good answer would be, "They are the numbers that are in one to one correspondence with the points on a line." Another might be, "They are the numbers which can be expressed by a decimal: terminating, repeating, or nonrepeating and non-terminating." Another answer is, "They are the numbers that are limits of 'convergent' sequences of rational numbers." Each of these answers should help you to feel better acquainted with the real numbers, but none of them gives a definitional description of the real number system.

The development of the real number system is a lengthy, involved undertaking. We would need to define a real number, then define equality, less than, addition, and multiplication, and develop the properties of the real number system. If this development were done in the same detail with provisions for experience that are provided in this book, it would take more pages than is expedient to include.

It is instructive and pertinent, however, to identify and relate an extremely important real number property with limits of sequences. Each treatment of number systems that presents a comparison of rational and real numbers will mention the **completeness property** of the real number system that does not hold in the rational number system. It can be stated in many forms:

(i) Every monotone nondecreasing sequence of real numbers that is bounded above has a limit which is a real number.

(ii) Every nonempty set of real numbers that is bounded above has a least upper bound. (By definition, a set S of real numbers is **bounded above** iff there is some real number N such that $s < N$ for every s in S. Similarly, the **least upper bound** of a set S of real numbers is a number m iff m is an upper bound and $m \leq m'$ where m' is any upper bound of S.)

(iii) If $\{(n, a_n)\}$ is a monotone nondecreasing sequence of real numbers and $\{(n, b_n)\}$ is a monotone nonincreasing sequence of real numbers such that both $a_n \leq b_n$ and $b_n - a_n < 1/10^n$ for every natural number n, then there exists a unique real number c such that $a_n \leq c \leq b_n$ for every natural number n.

Each of the statements listed above is equivalent; that is, given the algebraic and order properties of the real number system, the assumption of any one property is sufficient to prove the others. Indeed, there are still other statements for the completeness property that are equivalent to these, but many involve language not herein considered. The other statements and proofs of equivalence are to be found in mathematical literature.†

In every development of the real numbers, one will eventually encounter the association of an infinite numerical sequence with an unending decimal expression. For example, note the association of

$$0.3, 0.33, 0.333, \ldots, 0.333 \cdots 3, \ldots \text{ with } 0.3333\cdots,$$

and $\quad 0.2, 0.25, 0.250, \ldots, 0.250 \cdots 0, \ldots \text{ with } 0.2500\cdots.$

In general, if a is an integer and d_n, for every natural number n, is one of the digits 0, 1, 2, 3, 4, 5, 6, 7, 8, or 9, then

$$a.\, d_1,\ a.\, d_1 d_2, \ldots, a.\, d_1 d_2 d_3 d_4 \cdots d_n, \ldots$$

is associated with

$$a.\, d_1 d_2 d_3 d_4 d_5 d_6 \cdots.$$

The association is very natural because the unending decimal expression represents the **infinite series**

---

† See texts by Buck and Olmsted listed in bibliography.

$$a + \frac{d_1}{10} + \frac{d_2}{10^2} + \cdots + \frac{d_n}{10^n} + \cdots.$$

If you were told nothing more about such a series but instructed to evaluate it, you would probably begin to add the successive terms of the series and record your sums as follows:

$$a + \frac{d_1}{10}, a + \frac{d_1}{10} + \frac{d_2}{10^2}, \ldots, a + \frac{d_1}{10} + \frac{d_2}{10^2} + \cdots + \frac{d_n}{10^n}, \ldots.$$

This sequence is called a **sequence of partial sums** of the series.

What does an unending decimal expression represent? The completeness property, in particular statement (i), and the association between the unending decimal expression and the infinite numerical sequence enable us to give an answer: *Every unending decimal expression represents a real number.* For instance, consider the unending decimal expression

$$0.333333\cdots3\cdots$$

and an associated sequence s of partial sums

$$0.3, 0.33, 0.333, 0.3333, \ldots, 0.333\cdots3, \ldots.$$

The sequence s is monotone increasing, $s_{n+1} - s_n = 3/10^{n+1}$, and bounded above by 1. Therefore, it has a limit that is a real number. Also, notice that the limit of the sequence is $\frac{1}{3}$ (a result that you will verify in Exercise 4). Hence, we have an expected result using the completeness property; that is, $0.33333 \cdots$ represents the real number $\frac{1}{3}$.

Consider the unending decimal expression for $\sqrt{2}$,

$$1.414214\cdots.$$

The sequence of partial sums,

$$1.4, 1.41, 1.414, 1.4142, 1.41421, \ldots, \tag{1}$$

can be described as the sequence whose nth term is the largest number expressed with n decimal places and whose square is less than 2. Take for example the third term 1.414; it is expressed with three decimal places. Note that $1.414^2 = 1.999396 < 2$, but $1.415^2 = 2.002325 > 2$. This sequence is monotone nondecreasing and bounded above. Therefore, by statement (i) of the completeness property, the sequence has a limit which is a real number. The limit is $\sqrt{2}$ (as you can verify in the Exercises) and can be expressed by the infinite decimal $1.414214\cdots$.

In general, if we are thinking of an unending decimal expression

$$a. d_1d_2d_3d_4d_5d_6 \cdots d_n \cdots, \tag{2}$$

we are able to "evaluate" it or associate it with a sequence s of partial sums

$$a. d_1, a. d_1d_2, a. d_1d_2d_3, \ldots, a. d_1d_2d_3 \cdots d_n, \ldots. \tag{3}$$

This sequence s is monotone nondecreasing because $s_{n+1} - s_n = d_{n+1}/10^{n+1}$;

since d_{n+1} was defined to be a digit, $d_{n+1}/10^{n+1} \geq 0$. Also, this sequence s is bounded above by $a + 1$. Therefore, by statement (i) of the completeness property the limit of the sequence s exists and is a real number. We may not have a simple name such as $\frac{1}{3}$ or $\sqrt{2}$ for the limit of every sequence as represented in (3) but we always have at least one name available; that name is the unending decimal expression in (2).

Exercises

1. **(a)** Form a sequence of partial sums for the unending decimal expression $1.0000 \cdots$. **(b)** Find the limit of this sequence.

2. What unending decimal would you give for the product $3 \times 0.3333 \cdots$?

3. **(a)** Form a sequence s of partial sums for the unending decimal expression $0.9999 \cdots$. **(b)** Find the limit of this sequence.

4. Prove that the limit of the sequence $t = 0.3, \ 0.33, \ \ldots, \ 0.333 \cdots 3, \ \ldots$ is $\frac{1}{3}$. *Hint:* Notice that $t_n = \dfrac{333 \cdots 3}{10^n}$ and that $\left| \dfrac{333 \cdots 3}{10^n} - \dfrac{1}{3} \right|$ can be simplified by subtracting fractions. You may want to use the property that $10^n > n$ for every natural number n.

For Exercises 5 through 8 let the sequence $r = 1.4, \ 1.41, \ 1.414, \ \ldots$ be the sequence of partial sums whose nth term is the largest number expressed with n decimal places whose square is less than 2 [see (1)].

5. By the definition of the sequence r, what is the relationship between r_n and $\sqrt{2}$ for every n?

6. What is the relationship between $r_n + \dfrac{1}{10^n}$ and $\sqrt{2}$ for every n?

7. If $\left(r_n + \dfrac{1}{10^n} \right) - r_n = \dfrac{1}{10^n}$ what can be said concerning the closeness of r_n to $\sqrt{2}$; that is, about $|r_n - \sqrt{2}|$?

8. To prove that $\lim r_n = \sqrt{2}$ let any $\epsilon > 0$ be given, and then select $\delta = 1/\epsilon$. If $n > \delta$ and $\delta = 1/\epsilon$, then $10^n > 1/\epsilon$. Now verify that

$$|r_n - \sqrt{2}| < \epsilon \quad \text{if} \quad n > \delta.$$

chapter 2

Functions and Their Limits

It has long been the quest of inquiring minds to determine functional relationships between certain entities: the planets, stars, and earth; the price of stocks and the time to buy or sell; or heredity, environment, and an individual. These and a multitude of others that could have been named are very complex relationships. They have been studied and will continue to be investigated by the isolation of some small facets and the consideration of the simpler, directly related component parts.

Mathematics, which has developed methods for analyzing certain types of functional relationships, has accomplished a great deal. It is our purpose in this chapter to become familiar with the most basic functional relationship, the function, and a vitally important concept in the analysis of functions, the limit.

2-1 Functions

In Chapter 1 we defined a sequence to be a function in which each natural number was associated with one and only one real number. Although we have not defined a function in general, our procedure was pedagogically sound because most readers already had an entrenched idea of a sequence. Now we extend our concept of a function.

Since the domain of a sequence is the set of natural numbers, we frequently express the sequence by listing the range elements. For example, from the expression $1, \frac{1}{2}, \frac{1}{3}, \frac{1}{4}, \ldots, 1/n, \ldots$ we know that 1 is associated with 1, 2 is associated with $\frac{1}{2}$ (3 with $\frac{1}{3}$, 4 with $\frac{1}{4}$, etc.), and in general any natural number

n is associated with $1/n$. The set of associated numbers is a set of ordered pairs of real numbers. In our example this set is $\{(n, s_n)\}$ where n is any natural number and s_n, the associate of n, is $1/n$. The set $\{(n, s_n)\}$ includes $(1, 1)$, $(2, \frac{1}{2})$, $(3, \frac{1}{3})$, $(4, \frac{1}{4})$ and so forth. Knowledge of the set $\{(n, s_n)\}$ is necessary and sufficient for unique characterization of the sequence and recognition of this sequence among other sequences. This set of ordered pairs provides us with an example of the heart of the concept of a function.

To define a function we require two sets of numbers A and B and a third set f of ordered pairs of numbers (x, y) where x is an element of A and y is an element of B. Then f is a **function** iff for each x in A there is one and only one y in B such that (x, y) is in f. If the definition is satisfied we call the set A of all x the **domain of the function**, and the set of all y the **range of the function**. Notice that the set of all y is a subset, possibly a proper subset, of the set B. To determine specific functions, we must give the domain of the function and be able to identify the members of the set $\{(x, y)\}$. The range of the function can then be determined.

Example 1 Let sets A and B each be the set of real numbers. Then the set of ordered pairs of real numbers (x, y) such that $y = x^2$ is a function we shall call F. Give (a) the domain of F; (b) the range of F; (c) at least five elements of F.

> *Solution:* (a) The set A, the set of real numbers, is the domain of the function F. (b) The range of F is the set of non-negative real numbers—a proper subset of B. (c) The elements of F include $(1, 1)$, $(0, 0)$, $(2, 4)$, $(-\sqrt{2}, 2)$, $(-\frac{2}{3}, \frac{4}{9})$, and in general (x, x^2) for every real number x.

An **equation of a function** is an equation that expresses the relationship between the elements of the domain and the corresponding elements of the range. Two common equations for the function F in Example 1 are

$$y = x^2 \qquad \text{and} \qquad F(x) = x^2.$$

The symbol "$F(x)$" is convenient to represent the number associated with the number x by the function F.

Two functions f and g are **equal** or are called the *same* function iff their domains are the same set and $f(x) = g(x)$ for every real number x in the domain.

For convenience, we introduce the notation "D_f" to stand for the domain of a function f and the symbol "$\in$" which is read "is an element of." Thus "$x \in D_f$" is read "x is an element of the domain of f."

To identify a function f we shall indicate the domain D_f of the function f and some rule that may be used to associate one, and only one, real number

y with each real number x that is in D_f. For example, the function F in Example 1 may be identified as

$\qquad F(x) = x^2$, D_F is the set of real numbers,

$\qquad F = \{(x, y)\}$, $y = x^2$, D_F is the set of real numbers,

or $\qquad F = \{(x, x^2)\}$, D_F is the set of real numbers.

We shall find it convenient to agree that the domain of a function is the set of real numbers unless a statement is made to the contrary. Thus, the function F can be identified simply as

$$F = \{(x, x^2)\}.$$

Any function f may also be represented in *set builder notation:*

$$f = \{(x, y) \mid y = f(x), x \in D_f\}$$

which is read "f is the set of ordered pairs (x, y) such that $y = f(x)$ where x is an element of the domain of f." Again we should agree that D_f is the set of real numbers unless otherwise specified. When D_f is the set of real numbers we often write

$$f = \{(x, y) \mid y = f(x)\}$$

in general, or a statement of the form

$$\{(x, f(x)) \mid f(x) = x + 2\}$$

for a specific function.

Example 2 Let $G = \{(x, y) \mid y = 1/x, x \neq 0\}$. Notice that the statement $x \neq 0$ is to be used to identify the domain of the set of real numbers that are not 0. Give (**a**) $G(x)$; (**b**) the domain of G.

$\quad$ *Solution:* (**a**) For the function G, $G(x) = 1/x$. (**b**) The domain of G, D_G, is the set of all real numbers except 0.

$\quad$ The exercises below can help you to become better acquainted with the definitions and symbols that are provided in this text.

Exercises

1. Consider the function F, $F = \{(x, y) \mid y = x^2\}$, and evaluate: (**a**) $F(-3)$; (**b**) $F(3)$; (**c**) $F(5 - 7)$; and, where a and b are real numbers, (**d**) $F(a + b)$; (**e**) $F(a) + F(b)$; (**f**) $F(a \cdot b)$; (**g**) $F(a) \cdot F(b)$.

2. Consider the function $\{(x, G(x)) \mid G(x) = 3x\}$ and evaluate: (**a**) $G(-3)$; (**b**) $G(3)$; (**c**) $G(5 - 7)$; and, where a and b are real numbers, (**d**) $G(a + b)$; (**e**) $G(a) + G(b)$; (**f**) $G(a \cdot b)$; (**g**) $G(a) \cdot G(b)$.

3. Is it true for some, all, or no functions f that, if a and b are any two

elements in the domain, then (a) $f(a + b) = f(a) + f(b)$ or (b) $f(a \cdot b) = f(a) \cdot f(b)$?

4. A **constant function** is a function f in which all real numbers x in the domain of f are associated with exactly one real number c. Identify by an equation the constant function in which $c = 5$ and whose domain is the set of real numbers.

5. An **identity function** is a function f in which each real number x in the domain of f is associated with itself; that is, it consists of the ordered pairs (x, x) for each x in D_f. Identify in set builder notation the identity function whose domain is the set of real numbers.

2-2 Graphs of Functions

We define the graph of a function as we did the graph of a sequence: The **graph of a function** f is the set of all points on a coordinate system whose coordinates are $(x, f(x))$. We will use the Cartesian coordinate system exclusively to graph functions; the horizontal axis oriented to the right for the x values (domain) and the vertical axis oriented upward for the y values (range). Also, to avoid excessive wordage we may, if there is no chance for confusion, say "the point (x, y)" when we mean "the point whose coordinates are (x, y)."

As in the case of sequences the graph of a function can provide us with considerable insight and understanding of the function. Also, various terms that will be defined to classify functions originate from the graph of the function. We will analyze, compare, and classify functions in subsequent sections and, of course, we will determine various types of limits. However, let us now insure an acquaintance with the procedure of graphing functions by considering the following examples and exercises.

Example 1 Graph on a Cartesian coordinate system the function

$$f = \left\{ (x, y) \,|\, y = \frac{1}{x}; x > 0 \right\}.$$

Solution: The domain of the function is the set of positive real numbers, and the range is the same set because $1/x > 0$ if $x > 0$ and for any $y > 0$ there exists $x > 0$ such that $1/x = y$. Thinking of a few ordered pairs of the function will help us to get the general shape of the graph in mind:

$$\left(\frac{1}{2}, 2 \right), (1, 1), \left(2, \frac{1}{2} \right), \left(3, \frac{1}{3} \right), \dots, \left(x, \frac{1}{x} \right).$$

When we plot these points and mentally consider others, we are able to sketch the graph as shown in Figure 2–1.

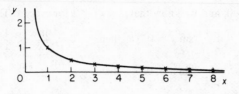

Figure 2-1

Example 2 Graph the function g given by the equation $y = \sin x$ whose domain is the set of positive real numbers. Notice that since x stands for a real number (or, if you wish, for the radian measure of an angle rather than the degree measure), the period of the sine function is 2π.

Solution: We will, of course, use a Cartesian coordinate system as in Example 1. If you are not acquainted with the sine function whose period is 2π, you should consult another book for the trigonometric functions of real numbers or of the radian measure of angles. In general, since an angle of x degrees has $\frac{1}{180}\pi x$ radians, the degree measure of an angle may be changed to radian measure, and vice versa. Accordingly, for $n = 0, 1, 2, 3, 4, \ldots$

$$\sin (n + 1)\pi = 0;$$

$$\sin \left(\frac{\pi}{2} + 2n\pi\right) = 1;$$

and $$\sin \left(\frac{3\pi}{2} + 2n\pi\right) = -1.$$

If you mentally consider values of x intermediate to these, then you should be able, with knowledge of trigonometry, to sketch the graph as in Figure 2–2.

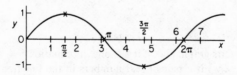

Figure 2-2

Example 3 Graph on a Cartesian coordinate system the function h whose equation is $y = \dfrac{\sin x}{x}$ and whose domain is $x > 0$. Use the same scale as in Examples 1 and 2 and note the relationship of this graph with the other two. This function is called a *damped sine function*.

Solution: Notice that since $\dfrac{\sin x}{x} = \dfrac{1}{x} \sin x$, we can evaluate $\dfrac{1}{x}$, evaluate $\sin x$, and then multiply these numbers to get $\dfrac{\sin x}{x} = y$. We are given $x > 0$, and we know that $|\sin x| \leq 1$ for every x. Thus, for $x > 0$

if $\sin x > 0$,   then $0 < \dfrac{\sin x}{x} \leq \dfrac{1}{x}$;

if $\sin x < 0$,   then $-\dfrac{1}{x} \leq \dfrac{\sin x}{x} < 0$;

if $\sin x = 0$,   then $\dfrac{\sin x}{x} = 0$;

if $\sin x = 1$,   then $\dfrac{\sin x}{x} = \dfrac{1}{x}$;   and

if $\sin x = -1$,   then $\dfrac{\sin x}{x} = -\dfrac{1}{x}$.

Thus, the graph of the function h whose equation is $y = \dfrac{\sin x}{x}$ will "oscillate" between the graph of f in Figure 2-1 and its reflection in the x-axis (whose equation is $y = -(1/x)$). Also, the graph of h will have the same x-intercepts as the graph of g in Figure 2-2. To exhibit the relationship more closely, we sketch the graphs of f, g, and the reflection of f lightly in broken lines and the graph of h in a dark solid line. (It is not in the realm of our discussion to determine the graph exactly when $0 < x < 1$; we will spotlight the graph at $x = 0$ in Example 3 of §2–13.)

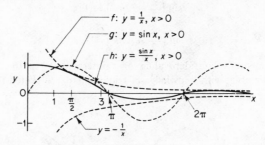

Figure 2-3

A function f, its graph, and, if it has one, its equation $y = f(x)$ are closely related since (x_0, y_0) is a pair of real numbers in the function iff (x_0, y_0) are coordinates of a point on the graph and iff the equation $y_0 = f(x_0)$ is true. Because of this close relationship you may frequently hear or say, "graph $y = f(x)$" when the graph of the function whose equation is $y = f(x)$ is wanted. In general, the **graph of an equation** in x and y is the set of all points on a coordinate system whose coordinates are the pairs of numbers (x, y) that satisfy the equation (i.e. the pairs of numbers in the truth set of the equation).

Example 4 Graph **(a)** the equation $x^2 + y^2 = 1$ and **(b)** the function $f = \{(x, y) \mid y = \sqrt{1 - x^2}, -1 \leq x \leq 1\}$.

Solution: In Figure 2–4 the required graphs are shown.

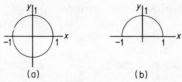

(a)   (b)

Figure 2-4

The following exercises should help you to become better acquainted with functions, their graphs, and their equations.

Exercises

Exercises 1 through 3 refer to the graphs of f, g, and h in Figure 2–3.

1. What points are held in common by the graphs of f and h?

2. For what positive values of x will $h(x) = 0$; that is, $(\sin x)/x = 0$?

3. Is there some real number $d > 0$ such that $(\sin x)/x = 0$ for every $x > d$?

4. Graph the constant function $\{(x, y) \mid y = 5\}$.

5. Graph the identity function $\{(x, y) \mid y = x\}$.

6. Compare the graphs of Exercises 4 and 5 with graphs on a Cartesian coordinate system of the sequences $\{(n, 5)\}$, and $\{(n, n)\}$.

7. Consider the graph of the equation $x^2 + y^2 = 1$ given in Example 4. **(a)** In this equation, if $x = 0$, what is y? **(b)** if $x = \sqrt{2}/2$, what is y? **(c)** Tell why this graph is not the graph of a function.

8. (a) Graph the function $g = \{(x, y) \mid y = -\sqrt{1 - x^2}, -1 < x < 1\}$. **(b)** Compare the graph of g with the graphs in Figure 2–4.

2-3 Some Examples to Ponder

The following functions are to be analyzed by you. You should graph each one and then use the graphs to compare the functions with each other and also to compare the functions of one group with those of another. You are probably expecting to find examples similar to the examples of sequences in § 1–3. If so, you are properly anticipating the extension of previously discovered concepts to this new topic, functions.

IMPORTANT NOTICE: Presently, our major concern will be with large positive values of x; you should focus your attention and concentration on the analysis of each function for these values of x (that is, at the right of the graph). In subsequent sections we will be primarily concerned with other values of x but with these same functions. Therefore, *the graphs that you make should be available for future reference.*

Group I	Group II	Group III
Ia. $\left\{\left(x,\frac{1}{x}\right)\right\}$, $x \neq 0$	IIa. $\left\{\left(x,\frac{1-x^2}{x}\right)\right\}$, $x \neq 0$	IIIa. $\{(x, \sin x)\}$
Ib. $\left\{\left(x,\frac{x-1}{x}\right)\right\}$, $x \neq 0$	IIb. $\left\{\left(x,\frac{x^2-1}{x}\right)\right\}$, $x \neq 0$	IIIb. $\{(x, 1 + \sin x)\}$
Ic. $\left\{\left(x,\frac{\sin x}{x}\right)\right\}$, $x \neq 0$	IIc. $\{(x, x)\}$,	IIIc. $\{(x, \cos x)\}$
Id. $\left\{\left(x,\frac{|x|}{x}\right)\right\}$, $x \neq 0$	IId. $\{(x, 2^x)\}$,	IIId. $\{(x, |\sin x|)\}$

If you have had trouble making any of the graphs that involve $\sin x$ or $\cos x$, consider the graph of those points for which $x = \frac{1}{2}n\pi$, where $n = 1, 2, 3, \ldots$. Then you can probably generalize from these specific points and sketch an adequate graph for all $x \geq \pi/2$.

If you have had trouble with the other graphs, consider the real numbers 1, 2, 3, 4, 5, 6 and graph the points that have these values for x. For any function f, the accurate graphing of these points and the general consideration of $f(x)$ for $0 < x < 1$, $f(x)$ for $1 < x < 2$, and so forth will generally yield an adequate sketch of the graph for $x > 0$.

You might have expected the factor $(-1)^x$ in the above examples because $(-1)^n$ was a factor in the expression of many sequences. The absence of the factor was, however, necessary because $(-1)^{1/2} = \sqrt{-1}$ and $\sqrt{-1}$ is not a real number. Likewise, $(-1)^{m/n}$ is not a real number if n is even and m is odd, and $(-1)^x$ is not defined when x is irrational. Therefore, $(-1)^x$ would not be a real number for some values of x, and it is not used for that reason.

In your classification of the examples you may have chosen to use names such as "increasing", "bounded", and so forth if they describe properties of functions that are similar to properties of sequences. You are urged to use such names in the following exercises—exercises that are designed to assist you in your classification.

Exercises

The following exercises pertain to the examples in Groups I, II, *and* III.

1. Identify the functions for which $f(x) \leq 2$ for every $x > 1$.

2. Identify the functions for which $f(x) \leq 1$ for every $x > 1$.

3. Identify the functions for which there is some real number d such that $f(x) \leq \frac{1}{10}$ for every $x > d$.

4. Identify the functions for which $f(x) \geq -1$ for every $x > 1$.

5. Identify the functions for which there is some real number d such that $f(x) \geq -\frac{1}{10}$ for every $x > d$.

6. What functions are included in the answers to Exercise 3 and Exercise 5.

7. Consider $x \geq 1$ only. Identify the functions for which there are real numbers M and N such that $M \leq f(x) \leq N$ for every x.

8. Consider $x > 0$. **(a)** Identify the functions for which there are numbers M and N such that $M \leq f(x) \leq N$ for every x. **(b)** Compare the answers of Exercises 7 and 8(a).

9. Consider $x > 1$ only. Identify the functions for which the following property is true for every pair of real numbers x_1, x_2: If $1 < x_1 < x_2$, then $f(x_1) < f(x_2)$.

10. Consider $x > 1$ only. Identify the functions for which the following property is true for every pair of real numbers x_1, x_2: If $1 < x_1 < x_2$, then $f(x_1) \geq f(x_2)$.

2-4 Classification of Functions

In the preceding exercises it should have been apparent that there is some concern about the values of x that are being considered. For example, the function Ia. in Group I, $\{(x, 1/x)\}\ x \neq 0$, might be called *bounded* for $x \geq 1$ because $0 < 1/x \leq 1$ if $x \geq 1$, but it certainly would not be for all $x > 0$ because for every positive real number M there exists a real number x where $0 < x < 1/M$ such that $1/x > M$. With infinite sequences we did not have this trouble because our domain was always the set of natural numbers. However, for functions whose domains are various subsets of the set of real numbers it seems imperative that we identify some subsets that we shall use frequently.

The following chart conveniently gives accepted names, symbols, and graphical representation on a number line for frequently used sets of real numbers. Let a and b be real numbers with $a < b$, and let the set in question be the set of all real numbers x satisfying the inequality in the first column:

Set	Name	Symbol	Graph
$a < x < b$	open interval	(a, b)	$\longmapsto$ a b
$a \leq x \leq b$	closed interval	$[a, b]$	a b
$x > a$	open right ray	$(a, \rightarrow)$	a
$x \geq a$	closed right ray	$[a, \rightarrow)$	a
$x < b$	open left ray	$(\leftarrow, b)$	b
$x \leq b$	closed left ray	$(\leftarrow, b]$	b

You have probably noted the use of the parenthesis in the symbols to indicate that the "end point" is not included in the set, and the use of the bracket to indicate that the "end point" is included in the set. You should also notice that a parenthesis is used adjacent to an arrow in the symbol for a ray because there is no largest nor smallest real number. Also, let it be understood that the use of either symbol (a, b) or $[a, b]$ implies that $a < b$.

Now that we have identified some sets of real numbers we can extend the definitions concerning monotone and bounded sequences to similar definitions for functions of real numbers on particular subsets of the domains of the functions. For instance, we define a function f to be **bounded above** on a subset S of the domain of f iff there exists a real number N such that $f(x) \leq N$ for every x in S. As with sequences, this number N is called an **upper bound** of f on S. You are given the opportunity in Exercise 1 to define **bounded below, lower bound, bounded, least upper bound,** and **greatest lower bound** for a function on a subset S of the domain of f.

Example 1 Show that the function $\left\{\left(x, \dfrac{x-1}{x}\right)\right\}$, $x \neq 0$ (Ib. of § 2–3) is bounded on $[1 , \rightarrow)$.

> *Solution:* If $x \geq 1$, then x is in the domain of the function. Notice that $\dfrac{x-1}{x} = 1 - \dfrac{1}{x}$; and if $x \geq 1$, then $0 < \dfrac{1}{x} \leq 1$ and $0 > -\dfrac{1}{x} \geq -1$. Therefore, by adding 1 and reversing the statement of the inequality we get $0 \leq 1 - \dfrac{1}{x} < 1$; hence, $\dfrac{x-1}{x}$ is bounded in $[1, \rightarrow)$ by an upper bound 1 and a lower bound 0.

A function f is **monotone increasing** on a subset S of the domain of f iff for every pair of real numbers x_1, x_2 in S, $x_1 < x_2$ implies $f(x_1) < f(x_2)$. But f is called **monotone nondecreasing** on S iff in the subset S, $x_1 < x_2$ implies $f(x_1) \leq f(x_2)$. You may define **monotone decreasing, monotone nonincreasing,** and **monotone** functions on a subset S of the domain of the function (see Exercise 1).

Example 2 Show that the function $\left\{\left(x, \dfrac{x-1}{x}\right)\right\}$, $x \neq 0$ (Ib. of § 2–3) is monotone increasing on $(0, \rightarrow)$.

> *Solution:* If $x > 0$, then x is in the domain of the function. Notice that $\dfrac{x-1}{x} = 1 - \dfrac{1}{x}$; and if $x_2 > x_1 > 0$, then $\dfrac{1}{x_1} > \dfrac{1}{x_2} > 0$. Hence, $\dfrac{-1}{x_1} < \dfrac{-1}{x_2} < 0$ and $1 - \dfrac{1}{x_1} < 1 - \dfrac{1}{x_2} < 1$. Therefore, if $0 < x_1 < x_2$, then $\dfrac{x_1 - 1}{x_1} < \dfrac{x_2 - 1}{x_2}$; the function is monotone increasing on $(0, \rightarrow)$.

Many other definitions could have been given because there are many

other properties of functions. In your analysis you might have determined
properties for which one of the following names would seem appropriate:
symmetric with respect to some line, symmetric with respect to some point,
concave upward, concave downward or convex upward, continuous, discon-
tinuous. You are invited to use these and other names to describe properties
of the functions that you analyze. They are not defined here because our
attention is focused on limits. You may notice that some of the names
describe the graph of the function or the range values of the function rather
than the function itself. The use of such words as "increasing," "symmetric," and "continuous" is not surprising if we were to look at some textbooks
of a few years ago. There we might find little distinction between the function
and its range, and we would realize that the notation $f(x)$ was used to stand
for both the function and the range number associated with x. Let us re-
emphasize, it is the function that determines the graph and the $f(x)$ values
in terms of x. The function, its domain, range, and graph, are interrelated
but different entities. Because this book is written to help you understand
mathematics and read other books, we will utilize the popular names for
properties of functions while continuing to distinguish between a function,
its range, and its graph.

Exercises

1. Let f be a function and S be a subset of the domain of f. Give definitions
   of the following terms for f on S that are analogous to the definitions of
   §1–4.
   (a) bounded below;   (b) lower bound;
   (c) bounded;   (d) least upper bound;
   (e) greatest lower bound;   (f) monotone decreasing;
   (g) monotone nonincreasing;   (h) monotone.

Exercises 2 through 5 refer to the examples in Groups I, II, and III of §2–3.

2. Identify the functions that are (a) bounded above on $[1, \rightarrow)$; (b)
   monotone nondecreasing on $[1, \rightarrow)$.

3. (a) Identify the functions that are in the answers to both (a) and (b) of
   Exercise 2. (b) Do these functions seem to have any other special common
   property?

4. Identify the functions that are (a) bounded below on $[1, \rightarrow)$; (b) monotone nonincreasing on $[1, \rightarrow)$.

5. (a) Identify the functions that are both monotone nonincreasing and
   bounded below on $[1, \rightarrow)$. (b) Do these functions seem to have any other
   special common property?

6. Prove that a function f is bounded on a subset S of the domain of f iff there exists a positive real number M such that $|f(x)| < M$ for every x in S.

2-5 A Limit of a Function: Informal

Have you sensed the presence of a limit for some functions? You have graphed and analyzed the function $\{(x, 1/x)\}$, $x \neq 0$ (Ia. of §2–3) and the sequence $\{(n, 1/n)\}$ (Exercise 1 of §1–9). How are these related? Is the function monotone and bounded on $(1, \rightarrow)$? Is the sequence monotone and bounded? What is the limit of the sequence? If the function has a limit, what should it be?

Compare the sequence $\left\{\left(n, \dfrac{n-1}{n}\right)\right\}$ (Ib. of §1–3) with the function $\left\{\left(x, \dfrac{x-1}{x}\right)\right\}$, $x \neq 0$ (Ib. of §2–3). If we considered the function on $[1, \rightarrow)$, then we could say that both the sequence and the function are monotone increasing and bounded. What is the limit of the sequence? If the function has a limit, what should it be?

The functions mentioned above are monotone on $(0, \rightarrow)$. Consider a function that is not monotone on any right ray, for example, $\left\{\left(x, \dfrac{\sin x}{x}\right)\right\}$, $x \neq 0$ (Ic. of §2–3). The graph of this function for $x > 0$ was shown in Figure 2–3 of §2–2. If the numbers x in the domain of this function were measures of time and $(\sin x)/x$ represented the distance of a swinging door from its ordinary rest position, would such a door ever come to rest? Of course, with friction there could not be such a perpetually oscillating free-swinging door. However, what about the function? Do you think that this function has a limit?

Consider the function $\left\{\left(x, \dfrac{|x|}{x}\right)\right\}$ on $(0, \rightarrow)$ (Id. of §2–3). How does it compare to the constant sequence $\{(n, 1)\}$? What would you expect the limit of this function to be?

As in Chapter 1, you should be a participant, not just a spectator, in identifying and defining a limit. The examples that you have analyzed were especially devised to enable you to recognize a particular type of limit of a function that is closely related to the limit of a sequence. If you have completed the first step and recognized some limits, now is the time for you to solidify your concept by forming a workable definition. If you have not, return to your analysis and graphs to formulate an idea of a limit of a function.

The following examples and exercises should increase your competence to define and recognize one type of limit of a function. You will probably find assistance and insight if you keep the graphs from §2–3 at hand.

Example 1 Consider the function $f = \left\{(x, y) \mid y = \dfrac{2}{x}, x \neq 0\right\}$. Find two different sets of values for x in $(0, \rightarrow)$ for which $f(x)$ is within $\dfrac{1}{10}$ of 0.

Solution: Since we are to focus our attention to the right of 0, let us seek values of x such that $0 < \dfrac{2}{x} < \dfrac{1}{10}$. This inequality is equivalent to $x > 20$; so we have one set $(20, \rightarrow)$ such that if x is in $(20, \rightarrow)$ then $0 < \dfrac{2}{x} < \dfrac{1}{10}$. For another set we can select any real number t greater than 20; if $x > t$, then $x > 20$, and $0 < \dfrac{2}{x} < \dfrac{1}{10}$.

Example 2 Consider the function $\{(x, \sin x)\}$ (IIIa. of §2–3). For what values of x in $(0, \rightarrow)$ is $\sin x$ within $\frac{1}{2}$ of 0?

Solution: Let us refer to the graph of the sine function for $x > 0$ that was given in Example 2 of §2–2. Also, let us draw two horizontal lines at $y = \frac{1}{2}$ and at $y = -\frac{1}{2}$; that is, an ϵ-band with $\epsilon = \frac{1}{2}$. Any point whose coordinates are $(x, \sin x)$ will lie in the ϵ-band iff $|\sin x| < \frac{1}{2}$. We draw the graph below and darken the x points on the x-axis for which $|\sin x| < \frac{1}{2}$. Since $\sin 30° = \sin \frac{1}{6}\pi = \frac{1}{2}$, the set of all values of $x > 0$ for which $|\sin x| < \frac{1}{2}$ is the collection of open intervals $(n\pi - \frac{1}{6}\pi, n\pi + \frac{1}{6}\pi)$ $n = 1, 2, 3, \ldots$ and the open interval $(0, \frac{1}{6}\pi)$.

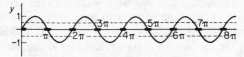

Figure 2-5

Exercises

Each of the following exercises refers to the examples in Groups I, II, *and* III *of* §2–3.

1. Consider Ia. $\{(x, 1/x)\}$, $x \neq 0$, on $(0, \rightarrow)$. **(a)** For what values of x is $|(1/x) - 0| < \frac{1}{2}$? **(b)** For what values of x is $|(1/x) - 0| < \frac{1}{10}$? **(c)** What is the greatest lower bound of $\{(x, 1/x)\}$, $x \neq 0$ on $[1, \rightarrow)$? **(d)** Do you think this function has a limit?

2. Consider Ic. $\left\{\left(x, \dfrac{\sin x}{x}\right)\right\}$, $x \neq 0$, on $(a, \rightarrow)$ where $a \geq 0$. **(a)** When $x = \dfrac{\pi}{2}, \dfrac{\sin x}{x} = \dfrac{2}{\pi}$. For what values of x is $\left|\dfrac{\sin x}{x} - 0\right| < \dfrac{2}{\pi}$? **(b)** When $x = \dfrac{5\pi}{2}, \dfrac{\sin x}{x} = \dfrac{2}{5\pi}$. For what values of x is $\left|\dfrac{\sin x}{x} - 0\right| < \dfrac{2}{5\pi}$? **(c)** Do you think this function has a limit?

3. Consider IIIa. $\{(x, \sin x)\}$, on $(0, \rightarrow)$ and review the results of Example 2. Is there any real number $d > 0$ such that $|\sin x - 0| < \frac{1}{2}$ for every $x > d$?

4. For what values of x in $(0, \rightarrow)$ are the range values of Ia. $\{(x, 1/x)\}$, $x \neq 0$, within ϵ of 0? (Remember, we agreed to use ϵ as a symbol for a positive real number.) *Hint:* Analyze the inequality $|(1/x) - 0| < \epsilon$.

5. For what values of x in $(0, \rightarrow)$ are the range values of Ib. $\left\{\left(x, \dfrac{x-1}{x}\right)\right\}$, $x \neq 0$, within ϵ of 1?

6. For what values of x in $(0, \rightarrow)$ are the range values of Id. $\left\{\left(x, \dfrac{|x|}{x}\right)\right\}$, $x \neq 0$, within ϵ of 1?

2-6 The Limit At-The-Right

Consider $\{(x, 1/x)\}$, $x > 0$. Is $1/x$ "close to zero" for any values of x in $(0, \rightarrow)$? Are there values of x such that $1/x$ is within $\frac{1}{10}$ of 0? Within $\frac{1}{100}$ of 0? Within ϵ of 0 where ϵ is any positive real number? The same type of set will answer each question above—what type of set? How do our questions and analysis compare with the questions and analysis of the sequence $\{(n, 1/n)\}$? If you have followed the development, answered the questions, and noted the association that has been made between certain sequences and functions, then you should be ready to complete the following sentence: By definition, the limit of the function $\{(x, 1/x)\}$, $x > 0$, is 0 because . . . .

If you wish the pleasure of *doing* mathematics rather than just *reading* about it, you should attempt to generalize and summarize your experiences to form a limit definition before reading further. Some statements follow that are good candidates for our definition. They are typical of statements that a good participating reader might give.

   (i) The number 0 is the limit of $\{(x, 1/x)\}$, $x > 0$, iff for every ϵ-band about 0 there is a right ray of x values such that the points with coordinates $(x, 1/x)$ are in the ϵ-band.

  (ii) The number L is the limit of the function f iff for every real number $\epsilon > 0$ there exists some real number d such that if $x > d$, $L - \epsilon < f(x) < L + \epsilon$.

 (iii) The number L is the limit of the function f iff $f(x)$ approaches L as x becomes infinite.

Notice that for (iii) the terms "approaches" and "becomes infinite" would need to be defined. Therefore, it is natural to expect to use phrases that involve previously defined terms or algebraic statements; as with the definition of the limit of a sequence, we want a precise language.

Each of the above statements includes the nucleus of the concept that has occupied our attention. However, each lacks complete consideration of all details that should be included in a definition. Primarily, recognize that our focus has been on the right hand side of the graph. Can there be another limit at the left? Certainly. Therefore, let us name the limit "at the right" to distinguish it from any other limit. Also, let us make sure that there are elements of $\{(x, f(x))\}$ "at the right" by insisting that the domain of f, D_f, includes some right ray and assuming that x is an element of D_f whenever we discuss $f(x)$. With these details incorporated, the following definition is given for later reference: Given a function f whose domain D_f includes an open right ray, then the real number L is the **limit at-the-right** of f iff for every real number $\epsilon > 0$ there exists a real number $\delta > 0$ such that

$$|f(x) - L| < \epsilon \quad \text{if} \quad x > \delta.$$

If the definition holds for a function f and a real number L, we shall write

$$\lim_{\rightarrow} \{(x, f(x))\} = L \qquad \text{or} \qquad \lim_{\rightarrow} f(x) = L.$$

Example 1 Let $f = \left\{(x, y) \mid y = \dfrac{2}{x},\ x \neq 0\right\}$. Prove, using the definition, that $\lim_{\rightarrow} (2/x) = 0$.

> *Solution:* The domain of f contains a right ray $(0, \rightarrow)$. Let any $\epsilon > 0$ be given. Then $0 < \dfrac{2}{x} < \epsilon$ iff $x > \dfrac{2}{\epsilon}$. Hence, for every $\epsilon > 0$ there exists $\delta > 0$ where $\delta = \dfrac{2}{\epsilon}$ such that $\left|\dfrac{2}{x} - 0\right| < \epsilon$ if $x > \delta$. Therefore by the definition of limit at-the-right, $\lim_{\rightarrow} \dfrac{2}{x} = 0$.

The task of proving by definition that $\lim_{\rightarrow} (2/x) = 0$ is similar to proving that $\lim_{\rightarrow} (2/n) = 0$. In general, if f is a function with domain D_f and we wish to use the definition to prove that $\lim_{\rightarrow} f(x) = L$, we do the following:

  (i) Insure that D_f contains an open right ray.
 (ii) Assume that ϵ is any given, fixed, positive real number.
(iii) Determine a fixed positive real number δ such that if $x > \delta$, then $|f(x) - L| < \epsilon$ (where δ is chosen to insure that $x \in D_f$ whenever $x > \delta$; otherwise, "$f(x)$" would have no meaning).

Because the definition requires a value of δ for each (every) value of ϵ, the positive number δ is frequently expressed in terms of ϵ. For instance, in Example 1 if $\epsilon = \frac{1}{10}$, we may take $\delta = 20$; if $\epsilon = \frac{1}{100}$, we may take $\delta = 200$; if any $\epsilon > 0$ is given, we may take $\delta = 2/\epsilon$ which is a specified, fixed real number associated with the given real number ϵ.

Example 2 Let $f(x) = x^2$, and consider the following argument for a claim that $\lim_{\rightarrow} x^2 = 0$: (i) $(0, \rightarrow)$ is in the domain of f. (ii) Let any $\epsilon > 0$ be given.

(iii) There exists $\delta = \dfrac{x^3}{\epsilon} > 0$ such that if $x > \dfrac{x^3}{\epsilon} > 0$, then $\dfrac{x^3}{x} < \epsilon$; that is,

$|x^2 - 0| < \epsilon$ if $x > \dfrac{x^3}{\epsilon} = \delta$. Hence, $\lim_{\rightarrow} x^2 = 0$.

Solution: Given $\epsilon > 0$, no fixed real number was named for δ; the expression (x^3/ϵ) is dependent upon x. For instance if $\epsilon = \frac{1}{10}$, $\delta = 10x^3$, but $10x^3$ is not a fixed real number—it depends on x. Actually, the function f is not bounded above on $(0, \rightarrow)$, for if N is any real number greater than 1, then $x^2 > N$ for every $x > N$. The argument is faulty; the function f has no limit at-the-right.

Exercises

Use the definition of limit at-the-right to complete the exercises. Assume the domain of each function f is the set of all real numbers x for which $f(x)$ is a real number.

1. Prove that $\lim_{\rightarrow} \{(x, 1/x)\} = 0$. Hint: Let ϵ be any positive real number and find a value for δ such that $|(1/x) - 0| < \epsilon$ if $x > \delta$.

2. Prove that $\lim_{\rightarrow} \dfrac{x-1}{x} = 1$. Hint: Remember that $\lim_{\rightarrow} \dfrac{x-1}{x}$ is the same as $\lim_{\rightarrow} \left\{ \left(x, \dfrac{x-1}{x} \right) \right\}$.

3. Prove that $\left\{ \left(x, \dfrac{|x|}{x} \right) \right\}$ has a limit at-the-right.

4. Prove that $\left\{ \left(x, \dfrac{1}{x-1} \right) \right\}$ has a limit at-the-right.

5. To prove that $\{(x, x)\}$ has no limit at-the-right, think of any real number L and some value of ϵ, say $\epsilon = 1$. (a) What is $|x - L|$ if $x > L + 1$? (b) Can there be any number δ such that $|x - L| < 1$ if $x > \delta$?

6. Prove that $\lim_{\rightarrow} \dfrac{\sin x}{x} = 0$

Exercises 7 and 8 are devised to assure you that it is the limit of a function that has been defined and not the limit of the set of $f(x)$ values. These two exercises provide examples of different functions with unequal limits at-the-right but functions with identical ranges.

7. Consider the two functions $\left\{ \left(x, \dfrac{1}{x} \right) \right\}$, $x > 1$, and $\left\{ \left(x, \dfrac{x-1}{x} \right) \right\}$, $x > 1$.
   (a) For the domain $x > 1$ as specified, what is the range of each function?
   (b) What is the limit at-the-right of each function?

8. Consider the two functions $\left\{ \left(x, \dfrac{|x|}{x} \right) \right\}$, $x \neq 0$, and $\left\{ \left(x, -\dfrac{|x|}{x} \right) \right\}$, $x \neq 0$.

(a) What is the range of each of these two functions? (b) What is the
limit at-the-right of each of these two functions?

2-7 The Limit At-The-Left

We have an excellent opportunity to extend the limit concept; let us capitalize
on our opportunity. To begin, let us recall the examples in Groups I, II, and
III of §2–3 and review or complete our analyses and graphs for $x < 0$. The
definitions in §2–4 of monotone and bounded functions apply to functions
on sets such as $(\leftarrow, 0)$ or $(\leftarrow, 1]$ as well as they did to functions on $[1, \rightarrow)$,
$(0, \rightarrow)$, and so forth.

Example 1 Classify the function $f = \{(x, 2^x)\}$ (IId. §2–3) on $(\leftarrow, 0)$ according to (a) boundness; (b) monotonicity.

> *Solution:* (a) If $x < 0$, then $2^x < 1$; and since $2^x > 0$ for every real
> number x, then $0 < 2^x < 1$. Hence, f is bounded on $(\leftarrow, 0)$. (b) If
> $x_1 < x_2 < 0$, then $2^{x_1} < 2^{x_2}$; hence, f is monotone increasing on $(\leftarrow, 0)$.
> (Someone might say, "f is monotone *decreasing as x decreases*." If this
> phrase were defined, this would be permissible. However, we need not
> and will not use the phrase in this book.)

What do you think would be the "limit at-the-left" of the examples in
Group I of §2–3? The limit at-the-right of $\left\{\left(x, \dfrac{|x|}{x}\right)\right\}$, $x \neq 0$, is 1. What do
you think the "limit-at-the-left" is? Of the examples in §2–3 there is one
that has no limit at-the-right but does have a "limit at-the-left." Which one
is it? If your answers to the last two questions were -1 and IId. respectively,
then you should try to organize your concept to form a new definition of
limit at-the-left in a manner similar to that previously instituted in forming
other definitions. This should be completed before reading further because
the definition follows: Given a function f whose domain D_f includes an
open left ray, then the real number L is the **limit at-the-left** of f iff for every
real number $\epsilon > 0$ there exists a real number $\delta > 0$ such that

$$|f(x) - L| < \epsilon \quad \text{if} \quad x < -\delta.$$

As was the case with the definition of limit at-the-right, we assume that δ is
chosen so that $x \in D_f$ whenever $x < -\delta$. If the definition holds for a function
f and a real number L, we shall write

$$\lim_{\leftarrow} \{(x, f(x))\} = L \quad \text{or} \quad \lim_{\leftarrow} f(x) = L.$$

Example 2 Prove that $\lim\limits_{\leftarrow} 2^x = 0$.

> *Solution:* We can refer to IId. of §2–3 where this function was first
> introduced. As is required in the definition, there is an open left ray
> $(\leftarrow, 0)$ in the domain of the function. Now we must show that for every

$\epsilon > 0$ there is a number $\delta > 0$ such that $|2^x - 0| < \epsilon$ if $x < -\delta$. If $\epsilon \geq 1$, then $\delta = 1 \, (-\delta = -1)$ would suffice because $|2^x - 0| = 2^x < 1 \leq \epsilon$ if $x < -1$. Our real concern is with "small values" of ϵ, $0 < \epsilon < 1$. Note that $|2^x| = 2^x$, and hence that $2^x < \epsilon$ iff $\log 2^x < \log \epsilon$. But $\log 2^x = x \log 2$. Therefore, $|2^x| < \epsilon$ iff $x \log 2 < \log \epsilon$; that is, iff $x < (\log \epsilon/\log 2)$. Thus, since $0 < \epsilon < 1$ for this case, $\log \epsilon$ and $(\log \epsilon/\log 2)$ are negative real numbers. Thus for $0 < \epsilon < 1$ we choose $\delta = - (\log \epsilon/\log 2)$. Therefore, $0 < \epsilon < 1$ and $\epsilon \geq 1$ have been successfully considered. Therefore, by definition, $\underset{\leftarrow}{lim} \, 2^x = 0$.

Example 3 Prove that $\{(x, 1 + \sin x)\}$ has no limit at-the-left.

Solution: The function $\{(x, 1 + \sin x)\}$ is IIIb. of §2–3. You should have a graph of the function for values of $x \geq 0$; now consider $x < 0$. Choose a value for ϵ, say $\epsilon = \frac{1}{2}$, and an ϵ-band about any suspected value of the limit L. Regardless of the choice of δ, there are smaller values of x, $x = (\pi/2) - 2n\pi < -\delta$ and $x = 3\pi/2 - 2n\pi < -\delta$ for some natural number n, such that $1 + \sin x = 2$ and $1 + \sin x = 0$, respectively. Notice that 2 and 0 cannot be within $\epsilon = \frac{1}{2}$ of any real number L; therefore, there can be no limit at-the-left for $\{(x, 1 + \sin x)\}$.

An investigation of older mathematics textbooks would probably yield different names for limit at-the-right and limit at-the-left. In regard to the limit at-the-right, we would probably find the clause "$f(x)$ approaches L as x becomes infinite" or "$f(x)$ approaches L as x increases without bound" and the symbol $\lim_{x \to \infty} f(x) = L$. If L is the limit at-the-left of a function f, we would find clauses such as "L is the limit as x becomes negatively infinite" or "as x decreases without bound" and the symbol $\lim_{x \to -\infty} f(x) = L$.

Indeed, such phrases as "$f(x)$ approaches," "x increases," and "x becomes" might intuitively be pleasing descriptions of the "action" taken by $f(x)$ relative to some "action" on the part of x. However, this "action" is totally dependent upon the function—upon the ordered pairs or functionally related pairs of numbers. We wish to define *limits of functions*. Therefore, the names "limit at-the-right" and "limit at-the-left" are given to the particular concepts that we developed because they are descriptive names that do not require any "action" by x or $f(x)$. If the definition of limit at-the-right holds for a function f and real number L, then f has a limit at-the-right—if the definition does not hold for any real number L, then f does not have a limit at-the-right. Regardless of what anyone claims x is "doing", the function does or does not have a limit at-the-right depending on whether or not the definition of limit at-the-right holds. We will neither use nor encourage usage of the words "approaches" or "becomes infinite" because we have not defined them and it would complicate rather than simplify our study to do so.

Exercises

Use the definition of limit at-the-left to complete Exercises 1 through 3. Assume the domain of each function f is the set of all real numbers x for which f(x) is a real number.

1. Prove that $\lim_{\leftarrow} \{(x, 1/x)\} = 0$.   **2.** Prove that $\lim_{\leftarrow} \dfrac{x-1}{x} = 1$.

3. Prove that $\left\{\left(x, \dfrac{|x|}{x}\right)\right\}$ has a limit at-the-left.

For Exercises 4 through 7, refer to the examples in Groups I, II, and III §2–3.

4. Identify those functions that are **(a)** bounded above on $(\leftarrow, 0)$; **(b)** monotone nonincreasing on $(\leftarrow, 0)$.

5. Identify those functions that are **(a)** bounded below on $(\leftarrow, 0)$; **(b)** monotone nondecreasing on $(\leftarrow, 0)$.

6. (a) What functions are monotone nonincreasing and bounded above on $(\leftarrow, 0)$? **(b)** Do these functions have a limit at-the-left?

7. (a) What functions are monotone nondecreasing and bounded below on $(\leftarrow, 0)$? **(b)** Do these functions have a limit at-the-left?

2-8 Theorems on Limits

In the preceding exercises you used the definitions to prove that certain functions have certain limits at-the-right or limits at-the-left. You assumed that ϵ was a positive real number and then found a positive value of δ, generally dependent upon ϵ, that enabled you to declare

$$|f(x) - L| < \epsilon \quad \text{if} \quad x > \delta$$
$$\text{or} \qquad |f(x) - L| < \epsilon \quad \text{if} \quad x < -\delta.$$

It would be worth your time to form some functions of your own that have a limit and to prove that the limit is what you think it is.

The functions in Groups II and III of §2–3 do not have a limit at-the-right, and, except for IId, they do not have limits at-the-left. To prove that a number L is not the limit at-the-right (or limit at-the-left) of a function f whose domain includes an open right ray (or open left ray), you should find some special value of ϵ such that, no matter what positive value is named for δ,

$$|f(x) - L| \geq \epsilon \quad \text{for some} \quad x > \delta.$$
$$\text{or} \qquad |f(x) - L| \geq \epsilon \quad \text{for some} \quad x < -\delta.$$

In addition to the direct application of these two limit definitions for functions, what theorems would you expect to be able to prove? The first limit theorem we proved for sequences was that the limit, if it exists, is a unique real number. If a function has one limit at-the-right (or at-the-left), L, could it have another, M? Reconsider Theorem 1–11a and its proof if you have any doubts about $L = M$; the proof that the limit at-the-right or at-the-left is unique is so like the proof of Theorem 1–11a that we do not repeat it here. We do, however, state the theorems for future reference.

Theorem 2-8a *Let f be a function whose domain D_f contains an open right ray. If $\lim_{\rightarrow} f(x) = L$ and $\lim_{\rightarrow} f(x) = M$, then $L = M$.*

Theorem 2-8b *Let f be a function whose domain D_f contains an open left ray. If $\lim_{\leftarrow} f(x) = L$ and $\lim_{\leftarrow} f(x) = M$, then $L = M$.*

If a function has a limit at-the-right, is the function bounded on some open right ray $(a, \rightarrow)$? This question suggests a theorem such as Theorem 1–11b, but wait. What about $\{(x, 1/x)\}$ on $(0, \rightarrow)$? This function has a limit at the right, 0, but it is not bounded on $(0, \rightarrow)$. However, notice that the function is bounded on $(1, \rightarrow)$. Consider the function $\left\{\left(x, \dfrac{x-1}{x}\right)\right\}$; it has a limit at-the-right. Is this function bounded on *some* open right ray? Even though it is not bounded on every open right ray [it is not bounded on $(0, \rightarrow)$] it is bounded on $(\frac{1}{2}, \rightarrow)$, $(1, \rightarrow)$, and many others.

You should consider other functions that have a limit at-the-right and determine if they are bounded on some open right ray. Also, you might consider functions that are not bounded on any open right ray and determine that they do not have a limit at-the-right. After working through some examples and organizing your thoughts you may anticipate the following theorem and proof.

Theorem 2-8c *If a function f has a limit at-the-right L, then f is bounded on some right ray.*

Proof: If $\lim_{\rightarrow} f(x) = L$, then the domain of f contains an open right ray; and for each real number $\epsilon > 0$ there exists a real number $\delta > 0$ such that

$$|f(x) - L| < \epsilon \quad \text{if} \quad x > \delta.$$

Choose $\epsilon = \frac{1}{2}$ and some $\delta > 0$ guaranteed by the definition, so that

$$|f(x) - L| < \tfrac{1}{2} \quad \text{if} \quad x > \delta;$$

that is, $\qquad L - \tfrac{1}{2} < f(x) < L + \tfrac{1}{2} \quad \text{if} \quad x > \delta.$

Hence, f is bounded on $(\delta, \rightarrow)$ by $L - \frac{1}{2}$ and $L + \frac{1}{2}$. The open right ray cited in the theorem is the open right ray $(\delta, \rightarrow)$.

A similar theorem and proof can be made for limit at-the-left. You are invited to form and prove such a theorem for yourself (see Exercise 7). Let us prove here a slightly different type of theorem:

Theorem 2-8d *If p is a positive real number and $\lim_{\rightarrow} f(x) = p$, then there exists some real number $\delta > 0$ such that $f(x) > \frac{1}{2}p$ if $x > \delta$.*

Proof: Since $\lim_{\rightarrow} f(x) = p$, $p > 0$, and $\frac{1}{2}p > 0$, then by the limit at-the-right definition, we are guaranteed some $\delta > 0$ such that

$$|f(x) - p| < \tfrac{1}{2}p \qquad \text{if} \qquad x > \delta.$$

But $|f(x) - p| < \frac{1}{2}p$ is equivalent to $p - \frac{1}{2}p < f(x) < p + \frac{1}{2}p$.

Therefore, $\qquad\qquad \frac{1}{2}p < f(x) \qquad \text{if} \qquad x > \delta.$

We have reached the desired conclusion.

As you probably suspect, similar theorems could be formed and proved for $\lim_{\rightarrow} f(x) = -p < 0$ and for limit at-the-left. Such tasks provide excellent opportunity to learn to prove limit theorems and become better acquainted with the limit definition; they are properly left as exercises (see Exercises 3 and 4).

In the definition of $\lim_{\rightarrow} f(x) = L$, we required that D_f contain an open right ray. Why was an *open* right ray required in the definition of limit at-the-right instead of *closed* right ray? This question and a similar one concerning left rays and the limit at-the-left should be answered. To be precise, we needed to state either open or closed. Also, in the limit at-the-right definition could we use $x \geq \delta$, instead of $x > \delta$? Exercises 1 and 2 are provided to enable you to understand that in either definition open or closed rays could have been used as could $x \geq \delta$ instead of $x > \delta$.

Exercises

Exercises 1 and 2 pertain to limit at-the-right. The same questions could have been asked and answered concerning a limit at-the-left.

1. Prove that a function f has an open right ray included in its domain iff there is a closed right ray included in its domain.

2. Let ϵ ($\epsilon > 0$) and L be real numbers. Prove there exists a positive real number δ such that $|f(x) - L| < \epsilon$ for every $x > \delta$ iff there exists a positive real number δ' such that $|f(x) - L| < \epsilon$ for every $x \geq \delta'$.

The exercises below are related to the exercises in §1–11 or to the theorems in this section.

3. Prove that if $p > 0$ and $\lim_{\rightarrow} f(x) = -p < 0$, then there exists a real number $\delta > 0$ such that $f(x) < -\frac{1}{2}p$ if $x > \delta$.

4. Prove that if $p > 0$ and $\underset{\leftarrow}{lim} f(x) = p > 0$ then there exists $\delta > 0$ such that $f(x) > \frac{1}{2}p$ if $x < -\delta$. (A similar statement and proof can be made by changing "$\underset{\leftarrow}{lim} f(x) = p$" to "$\underset{\leftarrow}{lim} f(x) = -p$.")

5. Consider the constant function, $\{(x, y)\}$, $y = c$ for every real number x. Prove that $\underset{\rightarrow}{lim} \{(x, c)\} = \underset{\leftarrow}{lim} \{(x, c)\} = c$.

6. Consider two functions f and g whose domains include $(a, \rightarrow)$ (where $a > 0$) such that $L \le g(x) \le f(x)$ for every x in $(a, \rightarrow)$. Prove that if $\underset{\rightarrow}{lim} f(x) = L$, then $\underset{\rightarrow}{lim} g(x) = L$. (A similar problem could be worded and proved for $(\leftarrow, a)$, $\underset{\leftarrow}{lim} f(x)$, and $\underset{\leftarrow}{lim} g(x)$. This statement is often treated as a theorem and is an application of the *domination principle* for limit at-the-right (or at-the-left) of a function.

7. State and prove a theorem for limit at-the-left that is analogous to Theorem 2–8c.

2-9 Another Type of Limit

The concept of a limit at-the-right or a limit at-the-left of a function is closely related to the limit of a sequence. The definitions of the three limits were of the form: for every $\epsilon > 0$ there exists $\delta > 0$ such that

if $n > \delta$ then $\quad |s_n - L| < \epsilon$ (for limit of a sequence),

if $x > \delta$ then $\quad |f(x) - L| < \epsilon$ (for limit at-the-right),

if $x < -\delta$ then $\quad |f(x) - L| < \epsilon$ (for limit at-the-left).

The graphical significance of the existence of a limit L was that we could locate points (n, s_n) or $(x, f(x))$ within any ϵ-band of L by choosing n "large enough" or x in some open ray. Also, if a function f has a limit at-the-left or limit at-the-right, then the graph of f has a horizontal asymptote.

There is another type of limit that is essential in the analysis of functions. If you have ever watched a televised rocket launching and heard a voice counting 5–4–3–2–1–0, you know that something very important will happen in an instant. With a function also, important results are frequently obtained as our attention is focused more and more sharply on some real number, say b. If our attention were on a graph of a function f on a Cartesian coordinate system, we would be carefully considering points of the graph that are close to the vertical line $x = b$.

Instead of telling you a definition outright, let us consider some examples and gain some experience to enable you to anticipate or form concepts for yourself. The two functions in Example 1 have the real number 0 extracted from their domain. Even when this is done, you can focus your attention on 0 and you can consider values of x on either side or both sides of 0. In the function g, you are to focus your attention on 0 and consider values of $x > 0$. Subsequently, we will turn our attention on other values of x and consider numbers on either side or both sides of these numbers, but at present 0 is adequate for illustration.

Example 1 Let $f = \left\{(x, y) \middle| y = \dfrac{x - 1}{x}, x \neq 0\right\}$ and $g = \{(x, y) \,|\, y = 1 - x,$ $x \neq 0\}$. Compare the analysis of f as you focus your attention on $(0, \rightarrow)$ at the right with the analysis of g as you consider $(0, \rightarrow)$ and focus your attention at 0.

Solution: We shall show a graph of each function and record our analysis in chart form beneath each graph.

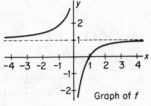

Figure 2-6

Graph of $f = \left\{\left(x, \dfrac{x - 1}{x}\right)\right\}$, $x \neq 0$.

Graph of $g = \{(x, 1 - x)\}$, $x \neq 0$.

On $(0, \rightarrow)$, f is monotone increasing and bounded above by 1.

On $(0, \rightarrow)$, g is monotone decreasing and bounded above by 1.

If $x > 10 > 0$, then

$$\left|\left(\frac{x - 1}{x}\right) - 1\right| = \left|\frac{1}{x}\right| < \frac{1}{10}.$$

If $0 < x < \dfrac{1}{10}$, then

$$|(1 - x) - 1| = |x| < \frac{1}{10}.$$

If $x > 100 > 0$, then

$$\left|\left(\frac{x - 1}{x}\right) - 1\right| = \left|\frac{1}{x}\right| < \frac{1}{100}.$$

If $0 < x < \dfrac{1}{100}$, then

$$|(1 - x) - 1| = |x| < \frac{1}{100}.$$

Let any $\epsilon > 0$ be given. Then

$$|f(x) - 1| < \epsilon \quad \text{if} \quad x > \frac{1}{\epsilon} > 0;$$

$$\lim_{\rightarrow} f(x) = 1.$$

Let any $\epsilon > 0$ be given. Then

$$|g(x) - 1| < \epsilon \quad \text{if} \quad 0 < x < \epsilon;$$

limit????

If you do not note in Example 1 a very close analogy between the two functions f and g as we have compared them, then draw ϵ-bands about the line $y = 1$ when $\epsilon = \frac{1}{10}$ and $\epsilon = \frac{1}{100}$ for each graph; then consider those values of x for which $f(x)$ and $g(x)$ are in the ϵ-band. Exercise 4 calls for a similar comparison of these two functions f and g on $(\leftarrow, 0)$.

In order to give you another example to ponder, we present one that is well known in analysis:

Example 2 Graph on a Cartesian coordinate system the function

$$h = \{(x, y) \mid y = \sin (\pi/x),\ x \neq 0\}.$$

Solution: Let us first consider some specific values of x, plot some individual points, and then generalize to sketch the graph for other points. The table below expresses conveniently a set of pairs $(x, \sin (\pi/x))$ where n is a natural number.

x	2	1	$\dfrac{2}{3}$	$\dfrac{2}{4}$	$\dfrac{2}{5}$	$\dfrac{2}{6}$	$\dfrac{2}{7}$	$\dfrac{2}{8}$	...	$\dfrac{2}{4n}$	$\dfrac{2}{4n+1}$	$\dfrac{2}{4n+2}$	$\dfrac{2}{4n+3}$
$\dfrac{\pi}{x}$	$\dfrac{\pi}{2}$	π	$\dfrac{3\pi}{2}$	2π	$\dfrac{5\pi}{2}$	3π	$\dfrac{7\pi}{2}$	4π	...	$\dfrac{4n\pi}{2}$	$\dfrac{(4n+1)\pi}{2}$	$\dfrac{(4n+2)\pi}{2}$	$\dfrac{(4n+3)\pi}{2}$
$\sin\dfrac{\pi}{x}$	1	0	-1	0	1	0	-1	0	...	0	1	0	-1

Figure 2-7

If $x > 2$, $0 < (\pi/x) < (\pi/2)$ and $0 < \sin (\pi/x) < 1$. If $x > 6$, $0 < (\pi/x) < (\pi/6)$ and $0 < \sin (\pi/x) < \frac{1}{2}$. Hence, we should suspect that $\underset{\rightarrow}{lim} (\sin (\pi/x)) = 0$. (We prove in Example 4 of §2–13 that indeed $\underset{\rightarrow}{lim} (\sin (\pi/x)) = 0$.) The value of $\sin (\pi/x)$ will oscillate between the values 1 and -1 as values of x are selected intermediate to those in the above chart. The trigonometric identity $\sin x = -\sin (-x)$ enables us to sketch the graph for negative values of x.

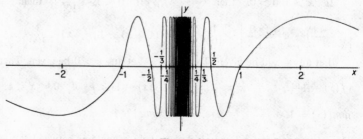

Figure 2-8

Exercises

Exercises 1 *through* 3 *pertain to Example* 2.

1. Consider the sequences $\left\{\left(n, \frac{2}{4n}\right)\right\}$ and $\left\{\left(n, \sin \frac{4n\pi}{2}\right)\right\}$.

    (a) What is $\lim_{\to} \frac{2}{4n}$?   **(b)** What is $\lim_{\to} \left(\sin \frac{4n\pi}{2}\right)$?

2. Consider the sequences $\left\{\left(n, \frac{2}{(4n+1)}\right)\right\}$ and $\left\{\left(n, \sin \frac{(4n+1)\pi}{2}\right)\right\}$.

    (a) What is $\lim_{\to} \frac{2}{(4n+1)}$?   **(b)** What is $\lim_{\to} \left(\sin \frac{(4n+1)\pi}{2}\right)$?

3. Select a positive real number d. **(a)** Is there a value of x in the open interval $(-d, d)$ such that $\sin \pi/x = 0$? **(b)** Is there a value of x in $(-d, d)$ such that $\sin \pi/x = 1$? **(c)** Could you select a smaller positive real number for d that would enable you to change your answers to questions (a) and (b)?

Exercises 4 *and* 5 *pertain to Example* 1.

4. Make an analysis of the functions in Example 1 for $(\leftarrow, 0)$ that parallels the one given for $(0, \rightarrow)$.

5. Do you think that g in Example 1 has a limit of any type?

2-10 More Examples to Ponder

If you have participated as suggested in the previous sections, you are probably anticipating a similar experience now. Also, if you have read and worked this far, you undoubtedly enjoy using your own mental powers to organize, analyze, offer and test conjectures, and draw conclusions for yourself. It would be unfair not to provide some examples that might enable you to define a new limit for yourself, but it would be somewhat unnecessary to burden you with many examples to classify according to boundedness, monotonicity, and so forth. Therefore, only a few specially chosen new examples are given in this section along with necessary directions; if you wish more functions to consider, recall those given in §2-3.

    First of all we "look at zero," and consider, "What happens to the functions there?" One answer might be, "Nothing—zero is not in the domain of any of the functions." This answer is true but incomplete; zero was purposely extracted from the domains to force you to determine distinguishing characteristics and compare the functions on the basis of those values of x "about zero" or "close to zero." However, it is zero and no other number "close to zero" or "far from zero" that is to be the focal point of your analysis; no other number is so strategically located that our answers would be the same.

Group I   Group II

Ia. $\left\{\left(x, \frac{x^2 + x}{x}\right)\right\}, x \neq 0$ IIa. $\left\{\left(x, \frac{|x|}{x}(x + 1)\right)\right\}, x \neq 0.$

Ib. $\{(x, |x|)\}, x \neq 0$ IIb. $\left\{\left(x, \frac{|x|}{x}\left[x + \frac{1}{10^6}\right]\right)\right\}, x \neq 0.$

Ic. $\{(x, y)\}, y = \begin{cases} x \text{ if } x \text{ is irrational,} \\ 0 \text{ if } x \text{ is nonzero rational.} \end{cases}$ IIc. $\{(x, y)\}, y = \begin{cases} 1 \text{ if } x \text{ is irrational,} \\ 0 \text{ if } x \text{ is nonzero rational.} \end{cases}$

If you have experienced any difficulties in the above examples this probably was related to one or more of the following: In IIa. and IIb. the factor $\frac{|x|}{x}$ is used to provide a factor 1 when x is positive and a factor -1 when x is negative. Both of these functions could have been expressed through compound sentences such as

$$\text{IIa. } \{(x, y)\}, y = \begin{cases} x + 1 & \text{if } x > 0, \\ -(x + 1) & \text{if } x < 0; \end{cases}$$

$$\text{IIb. } \{(x, y)\}, y = \begin{cases} x + \frac{1}{10^6} & \text{if } x > 0, \\ -\left(x + \frac{1}{10^6}\right) & \text{if } x < 0. \end{cases}$$

In Ia. the expression $\frac{x^2 + x}{x}$ is equivalent to $\frac{x + 1}{1}$ because $\frac{x}{x} = 1$ when $x \neq 0$; the expression $\frac{x^2 + x}{x}$ was used to stress the fact that 0 has been purposely extracted from the domain.

In Ic. and IIc. you are confronted with the impossible task of drawing an accurate graph. The rational numbers and the irrational numbers are both dense (between any two there is another), so that you have a dense set of points at $y = 0$ and another dense set at $y = 1$ or $y = x$. A dotted or dashed line has gaps of finite length, and hence does not have characteristics of a dense set. A straight solid line includes some points whose coordinates are rational and other points whose coordinates are irrational. What should you do? Sketch the graphs for Ic. and IIc. and mentally recognize and remember the characteristics of the graphs.

The following exercises should help you to classify the functions in Groups I and II. Feel free to ask and answer questions of your own invention, especially if they seem to be related to a limit concept.

Exercises

The following exercises pertain to the examples in Groups I *and* II *of this section.*

1. Let c be some positive real number. Classify the functions on the open interval $(0, c)$ according to: (a) monotonicity; (b) boundedness.

2. Choose a smaller positive value of c, say c'. How would the classification suggested for Exercise 1 be different on the interval $(0, c')$?

3. Let a be some negative real number. Classify the functions on the open interval $(a, 0)$ according to: (a) monotonicity; (b) boundedness.

4. Choose a larger negative value of a, say a'. How would the classifications suggested for Exercise 1 be different on the interval $(a', 0)$?

For Exercises 5 and 6, let a and c be real numbers where $0 < c < 1$ and $-1 < a < 0$.

5. Identify the least upper bound for each function that is (a) monotone decreasing and bounded above on $(0, c)$; (b) monotone increasing and bounded above on $(a, 0)$.

6. Identify the greatest lower bound for each function that is (a) monotone increasing and bounded below on $(0, c)$; (b) monotone decreasing and bounded below on $(a, 0)$.

7. The greatest lower bound of the function Ia. on the interval $(0, c)$ is 1 for every value of $c > 0$. (a) For what values of x on the interval $(0, 1)$ is $\dfrac{x^2 + x}{x}$ within $\frac{1}{10}$ of 1? (b) For what values of x on the interval $(0, 1)$ is $\dfrac{x^2 + x}{x}$ within ϵ of 1 where ϵ is some positive real number?

8. The greatest lower bound of the function IIa. on the interval $(0, c)$ is 1 for every value of $c > 0$. (a) For what value of x on the interval $(0, 1)$ is $\dfrac{|x|}{x}(x + 1)$ within $\frac{1}{10}$ of 1? (b) For what values of x on the interval $(0, 1)$ is $\dfrac{|x|}{x}(x + 1)$ within ϵ of 1 where ϵ is some positive real number?

2-11 Limits at a Real Number b

Consider the open interval $(0, 1)$ or an open interval $(0, c)$ where $c < 1$, and review Example 1 of §2–9 and the examples in Groups I and II of §2–10. Do any functions in these examples have a limit at 0? The function g in Example 1 of §2–9 and the real number 1 are related as follows: For every $\epsilon > 0$ there exists $\delta > 0$ $(\delta = \epsilon)$ such that $|g(x) - 1| < \epsilon$ if $0 < x < \delta$. This same ϵ,δ-statement would be true for the functions Ia. and IIa. in §2–10 if they were called g. Do any other of the examples in §2–10 have real numbers to which they are related as described above? Do any of the functions have a type of limit at 0?

The answers to the questions in the above paragraph are all affirmative provided we precisely describe the common characteristics under discussion and form a definition of the new kind of limit. You are urged to define for yourself the concept that has monopolized our attention; we shall call it the "right-side limit of a function at 0."

Although our attention has been focused at 0 to permit us to consider specific values, we can easily extend our concept to any real number b. The following definition should be a summary and natural extension of the limit concept that we have discovered: Given a function f whose domain D_f includes an open interval (b, c), then a real number L is the **right-side limit of f at** b iff for every real number $\epsilon > 0$ there exists a real number $\delta > 0$ such that

$$|f(x) - L| < \epsilon \quad \text{if} \quad 0 < x - b < \delta;$$

that is,

$$|f(x) - L| < \epsilon \quad \text{if} \quad x \in (b, b + \delta).$$

It is to be understood that δ is chosen so that $x \in D_f$ whenever $0 < x - b < \delta$ because we wish to discuss $f(x)$. If the definition holds for a function f and a real number L, then we shall write

$$\lim_{b^+} \{(x, f(x))\} = L \qquad \text{or} \qquad \lim_{b^+} f(x) = L.$$

Also, in other books you might find the symbol $\lim_{x \to b^+} f(x) = L$ or $\lim_{x \to b^+} f = L$.

Example 1 Consider Example Ib. of §2–10. Show that $\lim_{0^+} |x| = 0$.

> *Solution:* For every real number $c > 0$, the interval $(0, c)$ is in the domain of the function. Let $\epsilon > 0$ be given. Now, $||x| - 0| = |x| = x$ on $(0, c)$, and $||x| - 0| < \epsilon$ if $0 < x - 0 < \epsilon$. Since the interval $(0, \epsilon)$ is in the domain of the function, we can choose $\delta = \epsilon$; and hence, by the definition, $\lim_{0^+} |x| = 0$.

We shall find it convenient in Example 2 and in subsequent work to define a simple notation and introduce a symbol to be used. The **maximum** of a pair of real numbers consisting of δ_1 and δ_2 is δ_1 if $\delta_1 \geq \delta_2$, and is δ_2 if $\delta_1 < \delta_2$. We write max $[\delta_1, \delta_2]$ for the maximum of the pair δ_1, δ_2. In a similar manner, the **minimum** of the pair δ_1, δ_2 is δ_1 if $\delta_1 \leq \delta_2$ and is δ_2 if $\delta_1 > \delta_2$. We write min $[\delta_1, \delta_2]$ for the minimum of the pair δ_1, δ_2. Thus, max $[3, 5] = 5$; min $[3, 5] = 3$; max $[1, -9] = 1$; min $[1, -9] = -9$; max $[7, 7] = $ min $[7, 7] = 7$.

Example 2 Given $f = \{(x, x^2)\}$, determine $\lim_{1^+} x^2$.

> *Solution:* For every real number $c > 1$, the interval $(1, c)$ is in the domain of the function. Let $\epsilon > 0$ be given. If $x \in (1, c)$, then $|x^2 - 1| = |x + 1||x - 1| = (x + 1)(x - 1)$. Because we want a δ-condition on

$(x - 1)$, let us choose c, say $c = 2$, so that $0 < x + 1 < 3$ if $x \in (1, 2)$; that is, if $0 < x - 1 < 1$. Now, if $0 < x - 1 < 1$, then $(x + 1)(x - 1) < 3(x - 1)$; and if $0 < x - 1 < \frac{1}{3}\epsilon$, then $3(x - 1) < \epsilon$. Hence, choose $\delta = \min [\epsilon/3, 1]$, and then $(x + 1)(x - 1) < \epsilon$ if $0 < x - 1 < \delta$. Therefore, for every $\epsilon > 0$ there exists $\delta > 0$ ($\delta = \min [\epsilon/3, 1]$) such that $|x^2 - 1| < \epsilon$ if $0 < x - 1 < \delta$; that is, by definition, $\lim\limits_{1^+} x^2 = 1$.

The next definition that will be stated is that of the "left-side limit of a function at b." You probably can make the definition for yourself and you are urged to do so. Given a function f whose domain D_f includes an open interval (a, b), then a real number L is the **left-side limit of** f **at** b iff for every real number $\epsilon > 0$ there exists a real number $\delta > 0$ such that

$$|f(x) - L| < \epsilon \quad \text{if} \quad 0 < b - x < \delta;$$

that is,

$$|f(x) - L| < \epsilon \quad \text{if} \quad x \in (b - \delta, b).$$

Again it must be understood that δ is chosen so that $x \in D_f$ whenever $0 < b - x < \delta$ because we wish to discuss $f(x)$. If the definition holds for a function f and a real number L, then we write

$$\lim_{b^-} \{(x, f(x))\} = L \quad \text{or} \quad \lim_{b^-} f(x) = L.$$

Example 3 Consider Example Ib. of §2–10. Show that $\lim\limits_{0^-} |x| = 0$.

Solution: For every real number $a < 0$, the interval $(a, 0)$ is in the domain of the function. Let $\epsilon > 0$ be given. Now, $\big||x| - 0\big| = |x| = -x$ when x is a negative number. Hence, $\big||x| - 0\big| = -x < \epsilon$ if $0 < 0 - x < \epsilon$ or if $x \in (-\epsilon, 0)$. Because $(-\epsilon, 0)$ is in the domain of the function we can choose $\delta = \epsilon$; then $\big||x| - 0\big| < \epsilon$ if $0 < 0 - x < \delta$. Hence, by the definition, $\lim\limits_{0^-} |x| = 0$.

Example 4 Consider Example 2 of §2–9. Prove that $h = \{(x, y) \mid y = \sin (\pi/x), \ x \neq 0\}$ has no left-side limit at 0.

Solution: For every real number $a < 0$, the interval $(a, 0)$ is in the domain of h, and therefore this preliminary part of the definition holds. Now let $\epsilon = \frac{1}{4}$ be given, and assume that there exists some limit L and some positive real number δ such that $\sin (\pi/x)$ and L are within $\frac{1}{4}$ of each other whenever $0 < 0 - x < \delta$, as the definition requires. However, for every positive value of δ

$$\text{if} \quad n > \frac{1}{2\delta}, \qquad \text{then} \qquad 0 < 0 - \frac{-2}{4n} < \delta;$$

$$\text{and} \quad \text{if} \quad x = \frac{-2}{4n}, \qquad \text{then} \qquad \sin \frac{\pi}{x} = 0.$$

Also,   if $n > \dfrac{1}{2\delta} - \dfrac{3}{4}$,   then $0 < 0 - \dfrac{-2}{4n+3} < \delta$;

and   if $x = \dfrac{-2}{4n+3}$,   then $\sin \dfrac{\pi}{x} = 1$.

Note that 1 and 0 are not within $\frac{1}{4}$ of any real number L; hence, our assumption that there exists a limit L such that $|\sin(\pi/x) - L| < \frac{1}{4}$ if $0 < 0 - x < \delta$ is not always true. Hence, the definition fails, and $\lim\limits_{0^-}$ $\sin(\pi/x)$ does not exist.

The solution above involves some rather complicated reasoning. If we consider the graph of h (Figure 2–8 of §2–9) and imagine an ϵ-band with $\epsilon = \frac{1}{4}$ (a band $\frac{1}{2}$ unit in width), then we should probably say immediately that the left-side limit of h at 0 cannot exist. The graph does not lie within such an ϵ-band of any number for all values of x in any interval $(-\delta, 0)$. This is the same argument using the graph that we made in the above solution using algebra.

To conclude this section let us "look at both sides" of a real number b. If you reconsider the Examples to Ponder in §2–10, you should find that some have a left-side limit at 0 and some have a right-side limit at 0, and in some cases these two limits are equal. It is this latter type that we wish to characterize by defining a "limit at zero."

In general, if a function f has a left-side limit at b and a right-side limit at b and

$$\lim_{b^+} f(x) = \lim_{b^-} f(x) = L,$$

then we define L to be the **limit of f at** b and write

$$\lim_b \{(x, f(x))\} = L \qquad \text{or} \qquad \lim_b f(x) = L.$$

You probably anticipate other ways to denote this limit, such as $\lim\limits_{x \to b} f(x) = L$ or $\lim\limits_{x \to b} f = L$.

Example 5 Consider Example Ib. of §2–10 and Examples 1 and 3 of this section. Prove that $\lim\limits_{0} |x| = 0$.

> *Solution:* In Examples 1 and 3 we established that
>
> $$\lim_{0^+} |x| = 0 = \lim_{0^-} |x|.$$
>
> Hence, by definition, $\lim\limits_{0} |x| = 0$.

Example 6 Consider Example IIa. of §2–10. Determine if $\lim\limits_{0} \dfrac{|x|}{x}(x+1)$ exists.

Solution: The function is defined on the intervals $(a, 0)$ and $(0, c)$ whenever $a < 0 < c$. If $x > 0, \frac{|x|}{x}(x + 1) = x + 1$; and for every $\epsilon > 0, |(x + 1) - 1| < \epsilon$ if $0 < x < \epsilon = \delta$. Hence,

$$\lim_{0^+} \frac{|x|}{x}(x + 1) = 1.$$

If $x < 0, \frac{|x|}{x}(x+1) = -(x+1)$; and for every $\epsilon > 0, |-(x+1)-(-1)| = |-x| = -x < \epsilon$ if $0 < 0 - x < \epsilon = \delta$. Hence,

$$\lim_{0^-} \frac{|x|}{x}(x + 1) = -1.$$

Therefore, $\lim_{0^+} \frac{|x|}{x}(x + 1) \neq \lim_{0^-} \frac{|x|}{x}(x + 1)$; hence, $\lim_{0} \frac{|x|}{x}(x+1)$ does not exist.

While the definitions of the various limits at a real number b are fresh in your mind, let us return to the question, "What if $x = b$?" The answer is a nonrestrictive one: "So what—it makes no difference." A review of the definitions reveals that the condition $0 < x - b < \delta$ or $0 < b - x < \delta$ determines that we do not consider $x = b$. Therefore, the value of the function at b, and whether or not the function has a value at b have no bearing whatsoever on the existence of a left-side limit at b, right-side limit at b, or consequently a limit at b.

Exercises

Exercises 1 through 3 pertain to the examples in Groups I *and* II *of* §2–10.

1. Determine the right-side limit at 0, if it exists, of each function.

2. Determine the left-side limit at 0, if it exists, of each function.

3. Determine the limit at 0, if it exists, of each function.

4. Consider $f = \{(x, \sqrt{x})\}, x \geq 0$. Prove by the definition that $\lim_{0^+} \sqrt{x} = 0$.

5. Consider $\{(x, c)\}$, a constant function. Prove by the definition that $\lim_{b} c = c$ for every real number b.

6. Consider $H = \left\{\left(x, x\sin\frac{1}{x}\right)\right\} x \neq 0$. Prove that $\lim_{0}\left(x\sin\frac{1}{x}\right) = 0$. *Hint:* See Example 3 of § 2–2 for an analysis and graph of a similar function; the graph of *H* is "damped at 0" by $y = x$.

7. Let a be any real number. Prove that $\lim_{a}(mx + b) = ma + b$ where m and b are real numbers and $m \neq 0$.

8. Let b be any real number. Prove that $\lim_{b} x^2 = b^2$. *Hint:* Generalize Example 2.

2-12 More Theorems on Limits

There is a multitude of theorems that can be proved for limits of a function f at a real number b. If you have participated in the development of these latest limit definitions, you have probably noted some properties that you feel are valid consequences of our definition. You are invited to compile a list of conjectures that you have formed; that is, you are invited to list for yourself any property of a function that you feel is a sufficient condition or a necessary consequence for a left-side limit at b, a right-side limit at b, or a limit at b.

Now, trusting that you have listed your conjectures, you will recognize some of the following theorems. These are but a few of the theorems that could be proved. They are listed here because their proofs parallel proofs of similar theorems, or because they will be helpful to us in subsequent sections. You should not discard any conjecture of your own if it is not listed below; it may be among the theorems in the next chapter or you may, later and by yourself, prove the conjecture to be true or false.

Theorem 2-12a *If $\lim_{b^+} f(x) = L$ and $\lim_{b^+} f(x) = M$, then $L = M$.*

Theorem 2-12b *If $\lim_{b^-} f(x) = L$ and $\lim_{b^-} f(x) = M$, then $L = M$.*

These two theorems state that the right-side limit of a function f at b is a unique number, if it exists, and the left-side limit of a function f at b is a unique number, if it exists. The proof of these theorems depends on the same argument that was given in Chapter I for Theorem 1-11a. Only minor changes are required concerning δ and such phrases as $0 < x - b < \delta$ instead of $n > \delta$. Therefore, the proofs are left as exercises (see Exercise 1 and 2).

The validity of Theorem 2-12a and 2-12b enables us to conclude that the limit of a function f at b is unique if it exists.

Theorem 2-12c *If $\lim_{b} f(x) = L$ and $\lim_{b} f(x) = M$, then $L = M$.*

Proof: By definition, $\lim_{b} f(x) = \lim_{b^+} f(x) = \lim_{b^-} f(x)$. Therefore, if $\lim_{b} f(x) = L$ and $\lim_{b} f(x) = M$, then $\lim_{b^+} f(x) = L$ and $\lim_{b^+} f(x) = M$; and by Theorem 2-12a, $L = M$.

The definition of $\lim_{b} f(x)$ conveniently capitalized on the previously defined left-side and right-side limits at b. If the definitions of the right-side and left-side limits had been repeated, we would have had a definition that was different only in appearance. This is stated as a theorem even though the proof involves only a restatement of the definition.

Theorem 2-12d *Given a function f whose domain D_f contains open intervals (a, b) and (b, c), then a real number L is the limit of f at b iff for every real number $\epsilon > 0$ there exist positive real numbers δ_l and δ_r such that*

$$|f(x) - L| < \epsilon \quad if \quad 0 < b - x < \delta_l \quad or \quad 0 < x - b < \delta_r.$$

We shall find it convenient to introduce symbols and terminology regarding the condition $0 < b - x < \delta_l$ or $0 < x - b < \delta_r$. Note that

$$0 < b - x < \delta_l \quad \text{iff} \quad b - \delta_l < x < b; \quad \text{that is, iff} \quad x \in (b - \delta_l, b).$$

Also, $0 < x - b < \delta_r$ iff $b < x < b + \delta_r$; that is, iff $x \in (b, b + \delta_r)$.

The **union** of two sets A and B, written $A \cup B$, is the set of elements x such that $x \in A$ or $x \in B$. Hence, "$x \in (b - \delta_l, b)$ or $x \in (b, b + \delta_r)$" can be written "$x \in [(b - \delta_l, b) \cup (b, b + \delta_r)]$." An open interval (a, c) that contains b (i.e. $a < b < c$) is called a **neighborhood of** b. The union of two open intervals (a, b) and (b, c) is called a **deleted neighborhood of** b. Hence, $[(b - \delta_l, b) \cup (b, b + \delta_r)]$ is a deleted neighborhood of b. Furthermore, if $\delta_l = \delta_r = \delta$, the statement $x \in [(b - \delta_l, b) \cup (b, b + \delta_r)]$ is equivalent to the statement $0 < |x - b| < \delta$.

The convenient expressions involving deleted neighborhoods and the union of two sets enable us to state a necessary and sufficient condition for Theorem 2-12d and, consequently, for the definition of $\lim\limits_{b} f(x) = L$. Let $\lim\limits_{b} f(x) = L$; then, by Theorem 2-12d, D_f contains a deleted neighborhood of b, $[(a, b) \cup (b, c)]$, and for every $\epsilon > 0$ there exists $\delta_l > 0$ and $\delta_r > 0$ such that

$$|f(x) - L| < \epsilon \quad if \quad x \in [(b - \delta_l, b) \cup (b, b + \delta_r)]. \tag{1}$$

Choose $\delta = \min [\delta_l, \delta_r]$, and hence,

$$|f(x) - L| < \epsilon \quad if \quad x \in [(b - \delta, b) \cup (b, b + \delta)]; \tag{2}$$

that is, $|f(x) - L| < \epsilon$ if $0 < |x - b| < \delta$. $\tag{3}$

Conversely, for any $\epsilon > 0$, if there exists $\delta > 0$ such that statements (2) and (3) are true, then by choosing $\delta_l = \delta$ and $\delta_r = \delta$ statement (1) is true. Therefore we have proved a theorem that can be thought of as an alternate definition of the limit of a function at b.

Theorem 2-12e *Given a function f whose domain contains a deleted neighborhood of b, then a real number L is the limit of f at b iff for every $\epsilon > 0$ there exists $\delta > 0$ such that*

$$|f(x) - L| < \epsilon \quad if \quad x \in [(b - \delta, b) \cup (b, b + \delta)];$$

that is, $\quad |f(x) - L| < \epsilon \quad if \quad 0 < |x - b| < \delta.$

Now, if we have a function whose domain contains a deleted neighborhood of b, there are three ways to determine that the limit at b is or is not some real number L: Theorems 2–12d and 2–12e, and the definition of a limit at b. The following examples should assist us in retaining the elements of this discussion:

Example 1 Let $g = \{(x, \sqrt{x})\}$, $x \geq 0$. Prove that $\lim_{b} \sqrt{x} = \sqrt{b}$ providing $b > 0$.

Solution: Let any $\epsilon > 0$ be given. Select $\epsilon' = \min [\sqrt{b}, \epsilon]$; then

$$\epsilon' \leq \sqrt{b} \quad \text{and} \quad \sqrt{b} - \epsilon' \geq 0,$$

and $\quad |\sqrt{x} - \sqrt{b}| < \epsilon \quad$ whenever $\quad |\sqrt{x} - \sqrt{b}| < \epsilon'.$

Note that $\qquad\qquad\qquad |\sqrt{x} - \sqrt{b}| < \epsilon' \qquad\qquad\qquad\qquad$ (1)

iff $\qquad\qquad\qquad \sqrt{b} - \epsilon' < \sqrt{x} < \sqrt{b} + \epsilon'. \qquad\qquad\quad$ (2)

Since $\sqrt{b} - \epsilon' > 0$ (now you see why we selected ϵ' as we did) the inequality in (2) is equivalent to the inequalities

$$b - 2\epsilon'\sqrt{b} + \epsilon'^2 < x < b + 2\epsilon'\sqrt{b} + \epsilon'^2, \qquad\qquad (3)$$

and $\qquad -2\epsilon'\sqrt{b} + \epsilon'^2 < x - b < 2\epsilon'\sqrt{b} + \epsilon'^2. \qquad\quad$ (4)

Now we select $\delta_l = 2\epsilon' \sqrt{b} - \epsilon'^2 \quad (-\delta_l = -2\epsilon' \sqrt{b} + \epsilon'^2)$ and $\delta_r = 2\epsilon' \sqrt{b} + \epsilon'^2$. Thus if $0 < b - x < \delta_l$ or $0 < x - b < \delta_r$, then, by our choice of δ_l and δ_r, inequalities (4) through (1) hold and consequently $|\sqrt{x} - \sqrt{b}| < \epsilon$. Thus we have shown that for every $\epsilon > 0$ there exist $\delta_l > 0$ and $\delta_r > 0$ such that

$$|\sqrt{x} - \sqrt{b}| < \epsilon \quad \text{if} \quad 0 < b - x < \delta_l \quad \text{or if} \quad 0 < x - b < \delta_r.$$

Therefore, by Theorem 2–12d, $\lim_{b} \sqrt{x} = \sqrt{b}$ providing $b > 0$.

Example 2 and Figure 2–9 should enable us to relate the values given in the above solution and understand their association. Also, they will reacquaint us with the procedure of relating an algebraic analysis and the limit definition to the graph of a function.

Example 2 Sketch the graph of $g = \{(x, \sqrt{x})\}$, $x \neq 0$ and an ϵ-band about 2 with $\epsilon = 0.5$. Exhibit the relationship of the values δ_l, and δ_r which were determined in Example 1 for ϵ when $\epsilon = 0.5$, and exhibit a value of δ such that $|\sqrt{x} - 2| < 0.5$ if $0 < |x - 4| < \delta$.

Solution:

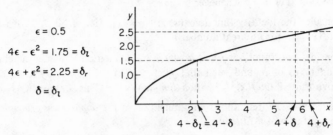

$\epsilon = 0.5$

$4\epsilon - \epsilon^2 = 1.75 = \delta_l$

$4\epsilon + \epsilon^2 = 2.25 = \delta_r$

$\delta = \delta_l$

Figure 2-9

In Figure 2–9, an ϵ-band about 2 is drawn such that a point $(x, f(x))$ is in the band iff $|\sqrt{x} - 2| < 0.5$. The vertical lines are specially located at 2.25 and 6.25 where if $x \in (2.25, 6.25)$ (including $x = 4$ in this example) then $1.5 < \sqrt{x} < 2.5$. The set of points between the vertical lines at $4 - \delta$ and $4 + \delta$ is a δ-**band** about 4 with the property that if $(x, \sqrt{x})$ is in the δ-band about 4, then $(x, \sqrt{x})$ is in the ϵ-band about 2. Notice that $4 - \delta = 2.25$ if $\delta = \delta_l$, but $4 + \delta = 5.75$ and $4 + \delta < 4 + \delta_r$ when $\delta = \delta_l$. The results of our discussion can be generalized to yield similar results for other functions and for either a left-side limit or a right-side limit.

The final theorem of this section is an extension of Theorem 2–8c.

Theorem 2-12f *If $\lim\limits_{b} f(x) = L$, then f is bounded on some deleted neighborhood of b.*

Proof: If $\lim\limits_{b} f(x) = L$, there are numbers a and c such that intervals (a, b) and (b, c) are in the domain of f. Also, if $\epsilon = \frac{1}{2}$, then by Theorem 2–12e there exists $\delta > 0$ such that $|f(x) - L| < \frac{1}{2}$ if $0 < |x - b| < \delta$; that is, f is bounded below by $L - \frac{1}{2}$ and bounded above by $L + \frac{1}{2}$ on the deleted neighborhood $[(b - \delta, b) \cup (b, b + \delta)]$.

Exercises

1. Prove Theorem 2–12a.

2. Prove Theorem 2–12b.

3. Prove the following sequel to Theorem 2–12f: If a function f has a right-side limit at b equal to L, then f is bounded on some interval (b, c).

4. Let f and g be two functions such that $f(x) = g(x)$ for every x in an open interval (a, b). Prove that if $\lim\limits_{b^-} f(x) = L$, then $\lim\limits_{b^-} g(x) = L$.

5. Prove that if $\lim_{b} f(x) = p > 0$, then there exists a real number $\delta > 0$ such that $f(x) > \frac{1}{2}p$ whenever $0 < |x - b| < \delta$.

6. Consider the identity function $\{(x, y) \mid y = x\}$, IIc. of §2–3. Prove that $\lim_{b} f(x) = b$ for any real number b.

7. Consider two functions f and g which are defined on a deleted neighborhood (a, c) of b, and such that $f(x) \leq g(x) \leq L$ for every x in (a, c). Prove that if $\lim_{b} f(x) = L$, then $\lim_{b} g(x) = L$. (This statement is often proved as a theorem and is an application of the *domination principle* for the limit at b of a function.)

2-13 Some Special Limits

In the examples that have been presented, there are many limits that we have not determined. When we were studying the examples in Groups I, II, and III of §2–3, we determined limits at-the-right and limits at-the-left, but we did not discuss the limit of any of the examples at a real number b. Similarly, when we investigated the examples in Groups I and II of §2–10 we determined limits at 0 only. Also, examples have been presented that involved trigonometric functions about which comments were made, but serious consideration was postponed (see Example 3 of §2–2 and Example 2 of §2–9). We will now consider these limits.

Our first objective is to evaluate

$$\lim_{0} \frac{\sin x}{x}.$$

To determine this limit a number of trigonometric inequalities must first be established and some limits of trigonometric functions must be evaluated. To determine a limit at 0 it is adequate to consider $-\frac{1}{2}\pi < x < \frac{1}{2}\pi$, $x \neq 0$. We shall exhibit a diagram and utilize trigonometric relationships and an intuitive geometrical argument to establish the needed equations and inequalities.

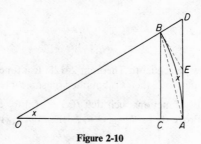

Figure 2-10

Let A and B be points of a unit circle with center at O such that $0 < x < \frac{1}{2}\pi$ where x is the radian measure of angle AOB. As in Figure 2–10 draw $\overline{BC}$ perpendicular to $\overline{OA}$ where C is on $\overline{OA}$, and draw $\overline{AD}$ perpendicular to $\overline{OA}$ where D is on OB. Since the radius of the circle is 1 and the ratio of the measures of two arcs is equal to the ratio of the measures of corresponding central angles, then

$$\frac{x}{2\pi} = \frac{\overparen{AB}}{2\pi(1)} \quad \text{and} \quad \overparen{AB} = x.$$

As in §1–5, we let "$\overline{AB}$" stand for the segment AB and also the length of $\overline{AB}$, and we let "$\overparen{AB}$" stand for the arc AB and also the length of $\overparen{AB}$.

Consider these statements for $0 < x < \frac{1}{2}\pi$:

$0 < \overline{BC}$ since B and C are distinct points;

$\overline{BC} = \sin x$ by the definition of $\sin x$;

$\overline{BC} < \overparen{AB}$ because $\overline{BC}$ is the perpendicular distance from
B to line OA and $\overparen{AB}$ is not.

Therefore, $0 < \sin x < x$ if $0 < x < \frac{1}{2}\pi$.   (1)

Next draw segment AB to form acute triangle AOB. Then $0 < \overline{AB} < x$ and by the law of cosines,

$$\overline{AB}^2 = 1^2 + 1^2 - 2\cos x,$$

$$x^2 > 2 - 2\cos x,$$

$$\frac{x^2}{2} > 1 - \cos x,$$

and $1 - \dfrac{x^2}{2} < \cos x.$

Therefore, if $0 < x < \frac{1}{2}\pi$, then $\cos x < 1$ and

$$1 - \frac{x^2}{2} < \cos x < 1 \quad \text{if} \quad 0 < x < \frac{1}{2}\pi.   (2)$$

Finally, draw $\overline{BE}$ tangent to circle O at B with E on $\overline{AD}$. Since two tangents $\overline{BE}$ and $\overline{EA}$ could be sides of a polygon circumscribed about circle O, then

$$\overline{AE} + \overline{BE} > \overparen{AB}.$$

Since $\overparen{AB} = x$

$$x < \overline{AE} + \overline{BE}.$$

Since triangle BED is a right triangle with right angle at B, then $\overline{BE} < \overline{ED}$ and

$$\overline{AE} + \overline{BE} < \overline{AE} + \overline{ED}.$$

But $\overline{AE} + \overline{ED} = \overline{AD}$ and $\overline{AD} = \tan x$. Therefore,

$$0 < x < \tan x \qquad \text{if} \qquad 0 < x < \tfrac{1}{2}\pi. \tag{3}$$

Now that the inequalities (1), (2), and (3) are established, we shall evaluate some helpful limits, $\lim_{0} \sin x$ and $\lim_{0} \cos x$, before determining $\lim_{0} \dfrac{\sin x}{x}$.

Example 1 Determine $\lim_{0} \sin x$.

> *Solution:* From our knowledge of the sine function we expect the limit to be 0. First, we note that the domain of the function contains a deleted neighborhood of 0. (Actually, 0 is also in the domain but this is irrelevant when finding the limit at 0.) We have established that
>
> $$0 < \sin x < x \qquad \text{if} \qquad 0 < x < \tfrac{1}{2}\pi, \text{ by (1)}.$$
>
> Also, we note that when $-\tfrac{1}{2}\pi < x < 0$, then $\sin(-x) = -\sin x$ and $0 < -x < \tfrac{1}{2}\pi$. Hence, from (1) we find
>
> $$0 < \sin(-x) < -x \qquad \text{if} \qquad -\tfrac{1}{2}\pi < x < 0$$
>
> which is equivalent to
>
> $$x < \sin x < 0 \qquad \text{if} \qquad -\tfrac{1}{2}\pi < x < 0. \tag{4}$$
>
> Thus, (1) and (4) permit us to state that
>
> $$|\sin x - 0| < |x| \qquad \text{if} \qquad 0 < |x| < \tfrac{1}{2}\pi.$$
>
> Now let $\epsilon > 0$ be given. Then $|\sin x - 0| < |x| < \epsilon$ if $0 < |x| < \epsilon$ and if (1) and (4) are true. Hence, choose $\delta = \min[\epsilon, \tfrac{1}{4}\pi]$, and then
>
> $$|\sin x - 0| < |x| < \epsilon \qquad \text{if} \qquad 0 < |x| < \delta.$$
>
> It follows then by the limit definition that $\lim_{0} \sin x = 0$.

Example 2 Determine that $\lim_{0} \cos x = 1$.

> *Solution:* From statement (2) we have
>
> $$1 - \frac{x^2}{2} < \cos x < 1 \qquad \text{if} \qquad 0 < x < \frac{1}{2}\pi.$$
>
> Also, since $0 < -x < \tfrac{1}{2}\pi$ whenever $-\tfrac{1}{2}\pi < x < 0$, it follows that
>
> $$1 - \frac{(-x)^2}{2} < \cos(-x) < 1 \qquad \text{if} \qquad -\frac{1}{2}\pi < x < 0. \tag{5}$$
>
> But, $(-x)^2 = x^2$ and $\cos(-x) = \cos x$; thus, (2) and (5) combine to yield
>
> $$1 - \frac{x^2}{2} < \cos x < 1 \qquad \text{if} \qquad 0 < |x| < \frac{1}{2}\pi.$$
>
> If we knew that $\lim_{0} \left(1 - \dfrac{x^2}{2}\right) = 1$, then the domination principle stated

in Exercise 7 of §2–12 would imply that $lim \cos x = 1$. It is your privi-
lege as a participating reader to complete this solution by proving
$lim_0 \left(1 - \dfrac{x^2}{2}\right) = 1$ (see Exercise 2).

Example 3 Determine $lim_0 \left(\dfrac{\sin x}{x}\right)$.

Solution: We will use the results of Example 2 and statements (1) and
(3). Statement (3) was

$$0 < x < \tan x = \frac{\sin x}{\cos x} \qquad \text{if} \qquad 0 < x < \frac{1}{2}\pi.$$

Multiplying the inequality in (3) by $\dfrac{\cos x}{x}$ we obtain

$$0 < \cos x < \frac{\sin x}{x} \qquad \text{if} \qquad 0 < x < \frac{1}{2}\pi.$$

But, since $\sin x < x$ if $0 < x < \frac{1}{2}\pi$ as in statement (1), then $\dfrac{\sin x}{x} < 1$

and $\qquad\qquad \cos x < \dfrac{\sin x}{x} < 1 \qquad$ if $\qquad 0 < x < \dfrac{1}{2}\pi.$ $\qquad$ (6)

If $-\frac{1}{2}\pi < x < 0$, then $0 < -x < \frac{1}{2}\pi$ and statement (6) yields

$$\cos(-x) < \frac{\sin(-x)}{-x} < 1 \qquad \text{if} \qquad -\frac{1}{2}\pi < x < 0. \qquad (7)$$

But $\cos(-x) = \cos x$ and $\dfrac{\sin(-x)}{-x} = \dfrac{\sin x}{x}$; thus (6) and (7) combine

to yield $\qquad\qquad \cos x < \dfrac{\sin x}{x} < 1 \qquad$ if $\qquad 0 < |x| < \dfrac{1}{2}\pi.$

Since $lim_0 \cos x = 1$, $lim_0 \dfrac{\sin x}{x} = 1$ by the domination principle stated in
Exercise 7 of §2–12.

Next let us turn our attention to $lim_{\to} \sin(\pi/x)$ that was considered in
Example 2 of §2–9. We have proposed that this limit is 0; now let us prove
that it is.

Example 4 Prove that $lim_{\to} \sin(\pi/x) = 0$, $x \neq 0$.

Solution: First note that the right ray $(0, \to)$ is in the domain of the
function. Now let any $\epsilon > 0$ be given. Since $(\pi/x) < (\pi/2)$ if $x > 2$, it
follows from (1) that

$$0 < \sin \frac{\pi}{x} < \frac{\pi}{x} \qquad \text{if} \qquad x > 2.$$

Furthermore, $(\pi/x) < \epsilon$ if $x > (\pi/\epsilon)$. Therefore, choose $\delta = \max$
$[(\pi/\epsilon), 2]$; it follows that

$$\left| \sin \frac{\pi}{x} \right| < \frac{\pi}{x} < \epsilon \quad \text{if} \quad x > \delta.$$

Thus, by definition, we conclude that $\lim_{\rightarrow} \sin (\pi/x) = 0$.

Exercises

1. Let t be a positive real number. Prove that $0 < x^2 < t$ if $|x| < \sqrt{t}$ (This statement should be helpful in completing Exercise 2).

2. Use the statement proved in Exercise 1 and the limit definition to prove that

$$\lim_{0} \left(1 - \frac{x^2}{2} \right) = 1.$$

3. Prove that $\lim_{\leftarrow} \sin (\pi/x) = 0$.

4. Prove that $\lim_{b} \dfrac{|x|}{x} = 1$ for every real number $b > 0$ (see Id. of §2–3).

5. Prove that $\lim_{0} (1 + \sin x) = 1$ (see IIIb. of §2–3).

6. Prove that $\lim_{0} 2^x = 1$ (see IId. of §2–3). *Hint:* Note that

$$1 - \epsilon < 2^x < 1 + \epsilon \quad \text{iff} \quad \epsilon < 1 \quad \text{and} \quad \frac{\log (1 - \epsilon)}{\log 2} < x < \frac{\log (1 + \epsilon)}{\log 2}.$$

7. Identify the examples, if there are any, in Groups I and II of §2–10 that have limits at-the-right or at-the-left.

2-14 Continuity

Closely related to the limit concept is the concept of *continuity*. Although the relationship may not be apparent to you at the outset we hope it will become obvious to you as you participate in the following developmental discussion. We begin with the assumption that you have some concept of continuity. Our purpose is to lead you from an intuitive concept to an appropriate mathematical definition through a discussion that primarily follows the historical development of continuity in mathematics.

To provide you an opportunity to sharpen your intuition and to provide us with specific examples to discuss, six familar functions are offered for consideration. They are named f, g, h, F, G, and H for convenient reference.

Example 1 Consider these functions and their graphs:

$$f = \{(x, x)\}, \qquad\qquad \text{(IIc. of § 2–3).}$$

$$g = \left\{ \left(x, \frac{|x|}{x} \right) \right\}, \, x \neq 0 \qquad \text{(Id. of § 2–3).}$$

$$h = \left\{\left(x, \frac{x^2 + x}{x}\right)\right\}, x \neq 0 \qquad \text{(Ia. of § 2–10)}.$$

$$F = \left\{\left(x, \frac{1}{x}\right)\right\}, x \neq 0 \qquad \text{(Ia. of § 2–3)}.$$

$$G = \left\{(x, y) \mid y = \frac{1}{x} \text{ if } x \neq 0, \text{ and } y = 1 \text{ if } x = 0\right\}.$$

$$H = \{(x, y) \mid y = x^2 \text{ if } x \neq 0, \text{ and } y = b \text{ if } x = 0\}.$$

(a) Identify each function whose graph can be drawn with an uninterrupted stroke of the pencil. **(b)** Discuss, intuitively, the continuity (or lack of it) of each graph. *As a reader, you will profit most from the following solutions if you analyze each graph and attempt to answer the questions for yourself.*

> *Solution:* **(a)** The graph of f can be drawn with an uninterrupted stroke of the pencil. Also, the graph of H can be sketched with an uninterrupted stroke *provided b is selected to be 0.*
>
> **(b)** The graphs of the functions have the following characteristics:
> f : continuous.
> g : not continuous; it has a two-unit gap at 0.
> h : not continuous; it has a hole at 0.
> F: not continuous; it diverges in opposite directions.
> G: not continuous; it diverges in opposite directions.
> H: depends on b—if $b = 0$ it is continuous; if $b \neq 0$ it is not continuous.

In Example 1 we relied on intuitive notions and vague phrases such as "gaps," "hole," and "diverges" to communicate an idea. Let us seek now to determine the concept of continuity, as it relates to functions, in terms of established mathematical concepts and a precise language.

Now is an appropriate time for you, the reader, to do some thinking with a pencil in your hand. The trouble spot of the functions discussed in Example 1 is at 0; hence, you can focus your attention there. How can continuous graphs be characterized? Why was $b = 0$ required for the graph of H to be continuous in Example 1? You should make an analysis of your own before reading Example 2.

Example 2 For each of the functions in Example 1 determine the answers to the following questions:
(a) Does the function have a value at 0?
(b) Does the function have a limit at 0?
(c) If $f(0)$ exists and $\lim_{0} f(x)$ exists, are they equal?
(d) Is the graph of the function continuous?

Solution: The questions are answered in Figure 2–11.

Question \ Function	(a) (value at 0)	(b) (limit at 0)	(c) ("value = limit")	(d) (continuous)
f	yes	yes	yes	yes
g	no	no	no limit or value	no
h	no	yes	no value	no
F	no	no	no limit or value	no
G	yes	no	no limit	no
H if $b = 0$	yes	no	yes	yes
H if $b \neq 0$	yes	yes	no	no

Figure 2-11

A close review of Figure 2–11 and of the previous discussion should enable you to anticipate a mathematically sound definition of what we might call "continuity at zero." In general, the *graph* of a function f is **continuous at b** iff

(i) $f(b)$ is defined,

(ii) $\lim_b f(x)$ exists, and

(iii) $f(b) = \lim_b f(x)$.

As was the case with other properties (monotone increasing, monotone decreasing, etc.) this definition is dependent exclusively upon properties of the function. Hence, we use continuity to describe graphs of functions, and we use continuity as a classification for functions. The following definition is widespread in mathematics: A *function f* is **continuous at b** iff

(i) $f(b)$ is defined,

(ii) $\lim_b f(x)$ exists, and

(iii) $f(b) = \lim_b f(x)$.

It follows immediately that a function f is continuous at b iff the graph of the function is continuous at b.

Our definition retains the intuitive notion that a graph which can be drawn with an uninterrupted stroke of the pencil is continuous. However, we have "received more than we bargained for." For instance, consider the next example.

Example 3 Determine whether the following function f is continuous at 0. (Note that f differs from Ic. of §2–10 because D_f contains 0.)

$$f = \{(x, y)\}, y = \begin{cases} x \text{ if } x \text{ is irrational,} \\ 0 \text{ if } x \text{ is rational,} \end{cases}$$

Solution: First, we note that $f(0) = 0$. Second, we note that $|f(x) - 0|$ $\leq |x - 0|$. Hence, for every $\epsilon > 0$ there exists $\delta > 0$, where $\delta = \epsilon$, such that $|f(x) - 0| < \epsilon$ if $0 < |x| < \delta$. Therefore, $\lim_{0} f(x) = 0$. Finally, we can conclude that this function called f and its graph are continuous at 0 by the definition, since

(i) $f(0) = 0$,

(ii) $\lim_{0} f(x) = 0$, and

(iii) $f(0) = \lim_{0} f(x) = 0$.

The very nature of the function and of rational and irrational numbers prevents us, however, from drawing a graph with a smooth stroke of the pencil.

To conclude this section, let us extend our concept to include more than continuity at a single real number b. The extension is quite natural: A function f (or the graph of a function f) is **continuous on a set** S of real numbers if f (or its graph) is continuous at each real number in S.

Example 4 Prove that $g = \{(x, \sqrt{x})\}$ is continuous on the set S of positive real numbers.

Solution: Let b be any positive real number. Then $g(b) = \sqrt{b}$ is a real number. We proved in Example 1 of §2–12 that $\lim_{b} g(x) = \sqrt{b}$ providing $b > 0$. Therefore, since (i) $g(b) = \sqrt{b}$, (ii) $\lim_{b} g(x)$ exists, and (iii) $\lim_{b} g(x) = \sqrt{b}$ for every real number $b > 0$, then g is continuous at b.

Exercises

For Exercises 1 through 4, prove that each function is continuous at 0.

1. $\{(x, \sin x)\}$ (IIIa. of §2–3). *Hint:* See Example 1 of §2–13.

2. $\{(x, 1 + \sin x)\}$ (IIIb. of §2–3). *Hint:* See Exercise 5 of §2–13.

3. $\{(x, \cos x)\}$ (IIIc. of §2–3).

4. $\{(x, x^2)\}$.

For Exercises 5 through 7, prove that each function is continuous on the set of all real numbers.

5. $\{(x, x)\}$ (IIc. of §2–3).

6. $\{(x, x^2)\}$. *Hint:* See Exercise 8 of §2–11.

7. $\{(x, c)\}$.

8. $\{(x, mx + b)\}$ where m and b are real numbers, $m \neq 0$.

9. Name a value for b (if there is any) for which the following function is continuous at 0:

$$\left\{(x, y) \mid y = \frac{x^2 + x}{x} \text{ if } x \neq 0, \text{ and } y = b \text{ if } x = 0\right\}.$$

10. Repeat Exercise 9 for $\{(x, y) \mid y = x \sin x \text{ if } x \neq 0, y = b \text{ if } x = 0\}$.

11. Repeat Exercise 9 for $\{(x, y) \mid y = (\sin x)/x \text{ if } x \neq 0, y = b \text{ if } x = 0\}$.

12. Repeat Exercise 9 for $\{(x, y) \mid y = \sin (\pi/x) \text{ if } x \neq 0, y = b \text{ if } x = 0\}$.

chapter 3

Generalization
and Application
of the Limit Concept

In Chapters 1 and 2 you became acquainted with some sequences and functions, you defined various limits, and you proved some theorems concerning each type of limit. Although you were actively involved in acquiring concepts and becoming acquainted with the definitions in the first two chapters, you probably asked yourself such questions as: "Is the tedious task of using the limit definition the only way to determine the limit of a function?" "Are there significant instances for the use of the limit concept other than those we have discussed (circumference of a circle, unending decimal expressions, continuity, etc.)?"

The purpose of this chapter is to provide you with adequate comprehension and appropriate tools which will enable you to determine efficiently many limits and which will give you a better understanding of the use of limits in mathematics. To accomplish this task we will determine ways of "breaking down" complicated functions, and we will generalize our limit definitions; then, with efficient effort we will prove theorems that will enable us to determine quickly (almost immediately) limits of a multitude of functions.

Let us initiate our ambitious undertaking without further ado.

3-1 Arithmetic of Functions

In order to evaluate limits of many functions we will find it convenient to start with arithmetic combinations of simpler functions. Then, as is often the case, we may find that some limits of seemingly complicated functions can be found from limits of simpler ones.

Even if you have not formally studied the arithmetic (or algebra) of functions, you are probably acquainted with the basic concepts involved. For example, consider the two functions called f and g:

$$f = \{(x, \sqrt{x})\}, \; x \geq 0; \text{ that is, } y = \sqrt{x}, \; x \geq 0.$$

$$g = \left\{\left(x, \frac{1}{x}\right)\right\}, \; x \neq 0; \text{ that is, } y = \frac{1}{x}, \; x \neq 0.$$

Although we neglect at present to mention a domain, a function is indicated by the set of ordered pairs

$$\left\{\left(x, \sqrt{x} + \frac{1}{x}\right)\right\} \text{ or the equation } y = \sqrt{x} + \frac{1}{x}.$$

What is the relationship of f and g to this indicated function? When the convention $f(x) = \sqrt{x}$ and $g(x) = 1/x$ is used, then the same function can be indicated by

$$\{(x, f(x) + g(x))\} \text{ or the equation } y = f(x) + g(x).$$

It now appears that the new function could be called the "sum of f and g" if a proper domain was expressed. Since we want both $f(x)$ and $g(x)$ to be real numbers, it should seem natural to select as domain the set of all real numbers that are in both the domain of f and the domain of g. Using D_f and D_g for the domains of f and g, respectively, we could say that we select $x \in D_f$ and $x \in D_g$.

The **intersection** $A \cap B$ of two sets A and B is the set of elements x such that x is an element of both A and B. Thus, the function under discussion can be precisely expressed as

$$\{(x, f(x) + g(x))\}, \; x \in D_f \cap D_g.$$

We could have discussed functions whose equations are

$$y = \sqrt{x} - \frac{1}{x}, \; y = \sqrt{x} \times \frac{1}{x}, \quad \text{and} \quad y = \sqrt{x} \div \frac{1}{x}$$

in a similar manner. In each case, the domain would be $D_f \cap D_g$ provided, of course, that we avoid having a zero divisor.

Now let us record the concepts discussed above by defining arithmetic combinations of functions. Let f and g be functions whose domains are D_f and D_g, respectively. The **sum**, $f + g$; **difference**, $f - g$; **product**, $f \times g$; and **quotient**, $f \div g$ of f and g are defined as follows:

Sum: $f + g = \{(x, f(x) + g(x))\}, \; D_{f+g} = D_f \cap D_g.$

Difference: $f - g = \{(x, f(x) - g(x))\}, \; D_{f-g} = D_f \cap D_g.$

Product: $f \times g = \{(x, f(x) \times g(x))\}, D_{f \times g} = D_f \cap D_g$.

Quotient: $f \div g = \{(x, f(x) \div g(x))\}, D_{f \div g} = D_f \cap D_g \cap Z$

where Z is the set of real numbers for which $g(x) \neq 0$.

If $x \in (D_f \cap D_g)$, then $x \in D_f$ and $x \in D_g$. Hence, both $f(x)$ and $g(x)$ are unique real numbers. Therefore, $f(x) + g(x)$, $f(x) - g(x)$, and $f(x) \times g(x)$ are unique real numbers. If $x \in (D_f \cap D_g \cap Z)$; then $x \in D_f$, $x \in D_g$, and $x \in Z$. Hence $f(x)$ and $g(x)$ are unique real numbers, and $x \in Z$ implies $g(x) \neq 0$. Therefore, $f(x) \div g(x)$ is a unique real number. Therefore, the sum, difference, product, or quotient of two functions is a function because a domain is specified, and with each real number x in the domain there is associated a unique real number.

It would be valuable at this point to study our definition, to create some examples, and to form and test some conjectures concerning algebraic properties of the functions. For instance, no order of operation needs to be specified in the sum $f + g$ and the product $f \times g$, but an order of operation is needed in the difference $f - g$ and the quotient $f \div g$. Why is this the case? What would $g + f$ and $g \times f$ be? By definition, $g + f = \{(x, g(x) + f(x))\}$ and the real number $g(x) + f(x)$ is the same as the real number $f(x) + g(x)$; hence, the functions $f + g$ and $g + f$ are the same. What about $f \times g$ and $g \times f$? You, the reader, may determine other properties of the sum or product of two functions, and you should consider some specific examples such as the following:

Example 1 Let $f = \{(x, x)\}$, $g = \{(x, 5)\}$, and $h = \{(x, |x|)\}$. Determine each of the functions: (a) $f \div h$; (b) $f \times (g + h)$; (c) $g \times (f \times f)$.

  Solution: (a) $f \div h = \left\{\left(x, \dfrac{x}{|x|}\right)\right\}$, $x \neq 0$. (b) $f \times (g + h) = \{(x, x(5 + |x|))\}$. (c) $g \times (f \times f) = \{(x, 5x^2)\}$.

Example 2 Let f, g, and h be defined as in Example 1. Express each of these functions as arithmetic combinations of f, g, and h: (a) $\left\{\left(x, \dfrac{|x|}{5}\right)\right\}$; (b) $\{(x, 25x)\}$; (c) $\left\{\left(x, \dfrac{|x|}{x}(x + 1)\right)\right\}$, $x \neq 0$.

  Solution: (a) $\left\{\left(x, \dfrac{|x|}{5}\right)\right\} = h \div g$. (b) $\{(x, 25x)\} = (g \times g) \times f$. (c) $\left\{\left(x, \dfrac{|x|}{x}(x + 1)\right)\right\}$, $x \neq 0$ is $(h \div f) \times (f + (g \div g))$.

Another important method of combining two functions is by *composition*. If f is the "squarer" function $\{(x, x^2)\}$ and g is $\{(x, x + 1)\}$ which we may call the "boost-one" function, then the composite function, f *of* g, is $\{(x, (x + 1)^2)\}$ which could be called the "square of the boost-one" function. The composite function, g *of* f, which we could describe verbally by the name "the boost-one of the square" function, would be $\{(x, x^2 + 1)\}$. Note that $f \, of \, g$ and $g \, of \, f$ are not the same functions.

A definition is in the offing and should be anticipated by you. A word of caution is pertinent, however, concerning the domain of a composite function. Informally and intuitively the problem should be formulated in the following questions: "If the square root function was working on the boost-one function, $\{(x, \sqrt{x + 1})\}$, what values could be used for x?" In precise terms, "If $F = \{(x, \sqrt{x})\}$, $x \geq 0$ and $G = \{(x, x + 1)\}$, what would you expect the domain of $\{(x, \sqrt{x + 1})\}$ to be?" Even though the domain of $\{(x, x + 1)\}$ is the set of real numbers, the domain of $(x, \sqrt{x + 1})$ could not contain any real number less than -1; we must give the domains special attention in our definition.

We define the **composite** of two functions as follows: Consider $f = \{(x, f(x))\}$, and $g = \{(x, g(x))\}$ with domain D_f and D_g, respectively; then

$$f \, of \, g = \{(x, f[g(x)])\} \, ; \text{ domain } D_{fofg} = \{x | x \in D_g \text{ and } g(x) \in D_f\}.†$$

If x is any number in D_{fofg}, then $x \in D_g$ and $g(x) \in D_f$. Since $x \in D_g$, $g(x)$ must be a unique real number in the domain of f and, hence, $f(g(x))$ is the unique real number associated with x. Therefore, the composite of two functions is a function because a domain is specified, and with each real number x in the domain there is associated a unique real number. As we have seen in our discussion, and shall show in the following example, the composite function $f \, of \, g$ need not be the same as the composite function $g \, of \, f$.

Example 3 Let $F = \{(x, \sqrt{x})\}$, $x \geq 0$ and $G = \{(x, x + 1)\}$. Find **(a)** $F(G(3))$; **(b)** $G(F(3))$. **(c)** Express $F \, of \, G$ in ordered pair notation. **(d)** Express $G \, of \, F$ in ordered pair notation.

> *Solution:* **(a)** $F(G(3)) = \sqrt{3 + 1} = 2$. **(b)** $G(F(3)) = \sqrt{3} + 1$. **(c)** $F \, of \, G = \{(x, \sqrt{x + 1})\}, x \geq -1$. **(d)** $G \, of \, F = \{(x, \sqrt{x} + 1)\}, x \geq 0$.

Example 4 Represent the following functions as composites of simpler functions: **(a)** $\{(x, \sin (x + 1))\}$; **(b)** $\{(x, 1 + \sin x)\}$; **(c)** $\{(x, \sin \sqrt{x})\}$, $x \geq 0$.

> *Solution:* Let $f = \{(x, \sin x)\}$, and $g = \{(x, x + 1)\}$. Then
> **(a)** $\{(x, \sin(x + 1))\}$ is $f \, of \, g$.   **(b)** $\{(x, 1 + \sin x)\}$ is $g \, of \, f$.
> **(c)** If $F = \{(x, \sqrt{x})\}$, $x \geq 0$, then $\{(x, \sin \sqrt{x})\}$, $x \geq 0$ is $f \, of \, F$.

---

† We are saying that to find the number associated with x in the function $f \, of \, g$, we first find the number associated with x in the function g, called, "$g(x)$," and then find the number associated with the number $g(x)$ in the function f, called "f of $g(x)$" and written "$f(g(x))$" or "$f[g(x)]$."

Exercises

1. Let two functions f and g be defined by their equations $y = f(x)$ and $y = g(x)$, respectively. Give the equations that define **(a)** $f + g$; **(b)** $f - g$; **(c)** $f \times g$; **(d)** $f \div g$; **(e)** f of g.

For Exercises 2 through 5 let $f = \{(x, \sqrt{x})\}$, $x \geq 0$; $g = \{(x, x + 1)\}$; $h = \{(x, |x|)\}$. Give ordered pair representation of each of these functions:

2. $f \times (f \times g)$.

3. $g - (f \div h)$.

4. f of h.

5. h of f.

For Exercises 6 through 9, represent the following functions in terms of the functions f, g, and h given for Exercises 2 through 5:

6. $y = (x + 1)^2$.

7. $y = \sqrt{(x + 1)^2}$.

8. $y = \sqrt{\sqrt{x}}$, $x \geq 0$.

9. $y = \sqrt{|x + 1|}$.

10. Let $f = \{(x, x)\}$, and let g be any function you choose. Determine **(a)** f of g and **(b)** g of f. (This function f is called the **identity function for composition.**)

For Exercises 11 through 14 give equations and domains to express the composite functions f of g and g of f where f and g are given in pairs.

11. $f(x) = x + 1$; $g(x) = x - 1$.

12. $f(x) = x^2$, $x \geq 0$; $g(x) = \sqrt{x}$, $x \geq 0$.

13. $f(x) = \log_{10} x$, $x > 0$; $g(x) = 10^x$.

14. $f(x) = \dfrac{1}{x}$, $x \neq 0$; $g(x) = \dfrac{1}{x}$, $x \neq 0$.

15. Two functions f and g such that f of $g = \{(x, x)\}$, $x \in D_{f \text{ of } g}$ and g of $f = \{(x, x)\}$, $x \in D_{g \text{ of } f}$ are called **inverse functions under composition**. Name four pairs of inverse functions under composition.

3-2 Interrelation of Limit Definitions

As we introduced new examples and developed definitions of various limits in Chapters 1 and 2, you probably noted a considerable amount of similarity between the definitions; in fact, we attempted to get you to extend your concepts from one definition to the next. Also, even the theorems that followed each limit definition were, in statement and proof, very closely related or analogous to those that followed other definitions. For instance, recall the theorems concerning the uniqueness of a limit:

Theorem 1–11a: If $\lim s_n = L$ and $\lim s_n = M$, then $L = M$.

Theorem 2–8 a: If $\lim_{\rightarrow} f(x) = L$ and $\lim_{\rightarrow} f(x) = M$, then $L = M$.

Theorem 2–8 b: If $\lim_{\leftarrow} f(x) = L$ and $\lim_{\leftarrow} f(x) = M$, then $L = M$.

Theorem 2–12a: If $\lim_{b^+} f(x) = L$ and $\lim_{b^+} f(x) = M$, then $L = M$.

Theorem 2–12b: If $\lim_{b^-} f(x) = L$ and $\lim_{b^-} f(x) = M$, then $L = M$.

Theorem 2–12c: If $\lim_{b} f(x) = L$ and $\lim_{b} f(x) = M$, then $L = M$.

It is both tiring and boring to read through this list of theorems when we all know that the "same" theorem was proved for each type of limit.

For another of many examples that could be given, consider the following theorems (which are paraphrased for emphasis):

Theorem 1–11b: If $\lim\ s_n = L$, then s is bounded.

Theorem 2–8 c: If $\lim_{\rightarrow} f(x) = L$, then f is bounded on some right ray.

Exercise 3, §2–12: If $\lim_{b^+} f(x) = L$, then f is bounded on an interval (b, c).

Could you add three more related theorems to this list? We have assumed that you could and that you would rather do so than have the theorems repeated for each of the other cases: $\lim_{\leftarrow} f(x)$, $\lim_{b^-} f(x)$, $\lim_{b} f(x)$. So alike are the limit definitions as we have given them that you should find the proofs of these theorems analogous to the proofs we presented.

We are faced with somewhat of a dilemma. We want to prove an adequate quantity of useful theorems concerning each type of limit. If we write out the theorem and proof for each type of limit our presentations will be bulky and boring to read; if we state and prove theorems for one or two types of limits only, our presentation will be incomplete. Since the definitions and many of the theorems seem to be very closely related, we may find a possible escape from our dilemma that would permit us to be both complete and concise. In this hope let us scrutinize our limit definitions and look for unifying general concepts.

Of the limit definitions in Chapters 1 and 2, two warrant special consideration: the limit of a sequence and the limit at b of a function. A sequence is a special type of function; where we previously called s_n the real number associated with n in the sequence s, let us name such a sequence f, use x in place of n, and $f(x)$ in place of s_n. The definition of limit at b was appreciably different, and for this limit we proved Theorem 2–12e, an alternate definition.

Since this alternate definition is similar in form to the other limit definitions, we will refer in this chapter to the statement in Theorem 2–12e as the definition of limit at b. Now, let us review and compare the limit definitions given in Chapters 1 and 2.

In the chart in Figure 3–1 we have an incomplete statement in which the blanks are labled with capital letters. Below the statement, six sets of completion items are listed horizontally and labeled I through VI. Insertion of the completion items of any set I through VI into the appropriate blanks will yield a paraphrasing of one of our six previously considered limit definitions.

$\boxed{\text{(A)}}$ $f(x) = L$ iff for every $\epsilon > 0$ there exists $\delta > 0$ such that

$|f(x) - L| < \epsilon$ if $\boxed{\text{(B)}}$.

	†(A)		(B)
I	Lim	(limit of a sequence)	$x > \delta$
II	Lim $\to$	(limit at-the-right of a function)	$x > \delta$
III	Lim $\leftarrow$	(limit at-the-left of a function)	$x < -\delta$
IV	Lim b^+	(right-side limit at b of a function)	$0 < x - b < \delta$
V	Lim b^-	(left-side limit at b of a function)	$0 < b - x < \delta$
VI	Lim b	(limit at b of a function)	$0 < |x - b| < \delta$

†A capital letter "L" is used in the abbreviation "Lim" at the beginning of a sentence.

Figure 3-1

You probably read only a short way through Figure 3–1 before you were able to anticipate subsequent entries. The six limit definitions give us six **types of limits**, as enumerated I through VI in Figure 3–1. Remember, however, that a function such as $f = \{(x, (1/x))\}$ has infinitely many particular limits: $\lim_{b} (1/x) = \lim_{b^+} (1/x) = \lim_{b^-} (1/x) = 1/b$ for infinitely many values of b. When we refer to a **particular limit**, let it be understood that we are considering one of the three types listed I, II, III in Figure 3–1; or, if we are considering one of the three types of limits "at b," then we have a particular real number b in mind. We shall use the symbol "$\lim$" to refer to the limit of a sequence, "$\lim_{b}$" to refer to the limit of a function at b, and the symbols "$\lim_{\to}$," "$\lim_{\leftarrow}$," "$\lim_{b^+}$," and "$\lim_{b^-}$" analogously.

Of course, our limit definitions in Figure 3–1 are not complete with all details. Therefore, let us list these necessary details because in our discussion of limits in general we shall assume that these items are understood. When we attempt to apply any theorems that we develop we will make sure that each particular problem satisfies each of these details.

(i) If a limit exists (call it L, M, or what you wish), it is a real number.

(ii) The function under discussion, call it f, is a function whose range is a subset of the set of real numbers.

(iii) If we consider $f(x)$ in a discussion involving a function f, then we insist that x is an element in the domain of the function f.

(iv) The domain D_f of the function f must be **appropriate** for the particular limit under consideration. These domains are reviewed in Figure 3–2.

Let a, b, c, p, q be real numbers with $a < b < c, p > 0$, and $q < 0$.

For	I,	$\lim$,	D_f is the set of natural numbers.
For	II,	$\lim_{\longrightarrow}$,	D_f contains an open right ray $(p, \longrightarrow)$.
For	III,	$\lim_{\longleftarrow}$,	D_f contains an open left ray $(\longleftarrow, q)$.
For	IV,	$\lim_{b^+}$,	D_f contains an open interval (b, c).
For	V,	$\lim_{b^-}$,	D_f contains an open interval (a, b).
For	VI,	$\lim_{b}$,	D_f contains a deleted neighborhood of b, such as
			$[(a, b) \cup (b, c)]$.

Figure 3-2

These details may seem so natural to you that you wonder why we listed them. We have done so just to ensure that we do not forget that they are necessary.

Now let us return our attention to the heart of the limit concept: the ϵ, δ-relationship. Each type of limit listed I through VI in Figure 3–1 is associated with certain sets of real numbers x such that x is in the domain of the function f, and x satisfies the inequalities in column (B). This association is apparent in the following chart:

In	I,	x satisfies	$x > \delta$	iff	$x \in (\delta, \longrightarrow)$.
In	II,	x satisfies	$x > \delta$	iff	$x \in (\delta, \longrightarrow)$.
In	III,	x satisfies	$x < -\delta$	iff	$x \in (\longleftarrow, -\delta)$.
In	IV,	x satisfies	$0 < x - b < \delta$	iff	$x \in (b, b + \delta)$.
In	V,	x satisfies	$0 < b - x < \delta$	iff	$x \in (b - \delta, b)$.
In	VI,	x satisfies	$0 < |x - b| < \delta$	iff	$x \in [(b - \delta, b) \cup (b, b + \delta)]$.

Figure 3-3

We call each of the sets listed in the right column of Figure 3–3, **δ-sets.** We use the Greek letter Δ, capital delta, which we call "del," to name a particular δ-set. For example, if δ_1 is a real number, the set of real numbers x such that $x \in (\delta_1, \longrightarrow)$ is Δ_1, called "del sub 1."

Can you give a single statement that will encompass the basic idea of the limit concept and that will apply to each particular limit we have discussed? The basic idea of the limit concept is the ϵ, δ-relationship. Can you express this relationship in a single statement that applies to each particular limit we have discussed? An answer to these questions is suggested in Figure 3–1 where we stated:

$\boxed{\text{(A)}}$ $f(x) = L$ iff for every $\epsilon > 0$ there exists $\delta > 0$ such that

$$|f(x) - L| < \epsilon \quad \text{if} \quad \boxed{\text{(B)}}.$$

Let us use the term "δ-set" to represent the entries that replace (B) and the word "*limit*" for the entries that replace (A). Hence, the statement,

> **Limit $f(x) = L$ iff for every $\epsilon > 0$ there exists a δ-set Δ such that**
> $$|f(x) - L| < \epsilon \quad \text{if} \quad x \in \Delta,$$

encompasses each of the limit definitions we have previously encountered. We will refer to this statement as the **generalized limit definition.**

When we apply the generalized limit definition to a specific function we shall be given some particular limit ($\lim_{\rightarrow}$, $\lim_{\leftarrow}$, $\lim_{b^+}$, $\lim_{b^-}$, $\lim_{b}$ or $\lim_{b}$). Then we shall make sure that f has an appropriate domain as described in Figure 3–2, and we shall determine for each given $\epsilon > 0$ some δ-set Δ that is contained in the domain of the function and such that

$$|f(x) - L| < \epsilon \quad \text{if} \quad x \in \Delta.$$

Example 1 Let $L = 0$, $f = \{(x, (1/x))\}$, $x \neq 0$. Show that the generalized limit definition applies to $\lim_{\rightarrow} (1/x) = 0$.

Solution: First we note that L is a real number, and f is a function whose domain contains an open right ray $(0, \rightarrow)$. Now let $\epsilon > 0$ be given. We note that

$$\left|\frac{1}{x} - 0\right| < \epsilon \quad \text{if} \quad x > \frac{1}{\epsilon}.$$

Hence, for any $\epsilon > 0$ we have exhibited a positive real number δ, where $\delta = (1/\epsilon)$ and, simultaneously, a δ-set Δ, where $\Delta = (1/\epsilon, \rightarrow)$, such that

$$\left|\frac{1}{x} - 0\right| < \epsilon \quad \text{if} \quad x > \frac{1}{\epsilon} \quad \text{and} \quad \left|\frac{1}{x} - 0\right| < \epsilon \quad \text{if} \quad x \in \Delta.$$

Thus if $f = \{(x, (1/x))\}$, $x \neq 0$ and $L = 0$, the generalized limit definition applies to $\lim_{\rightarrow}$ (limit at-the-right) to yield a statement that is equivalent to the definition of $\lim_{\rightarrow} (1/x) = 0$.

Example 2 Let $f = \{(x, (1/x))\}$, $x \neq 0$. Show that the generalized limit definition applies to $\lim_{b^+} (1/x) = (1/b)$, $b > 0$.

Solution: We are given b, a positive real number, and hence $1/b$ is a positive real number. Also, for each positive value of b, f is a function whose domain contains an open interval (b, c) for every real number c provided $c > b$. Let any $\epsilon > 0$ be given. Note that

$$\left|\frac{1}{x} - \frac{1}{b}\right| = \left|\frac{b - x}{xb}\right| = \left|\frac{x - b}{xb}\right| = \frac{x - b}{xb} \quad \text{if} \quad x > b > 0.$$

To handle the unwanted factor x in the denominator, we note that since $x > b$,

$$\frac{1}{x} < \frac{1}{b}, \quad \text{and} \quad \frac{x - b}{xb} < \frac{x - b}{b^2}.$$

Thus, $\dfrac{x - b}{xb} < \epsilon \quad$ if $\quad 0 < x - b < b^2\epsilon.$

Hence, for any $\epsilon > 0$ we have exhibited a positive real number δ, where $\delta = b^2\,\epsilon$, and, simultaneously, a δ-set Δ, where $\Delta = (b, b + b^2\,\epsilon)$, such that

$$\left|\frac{1}{x} - \frac{1}{b}\right| < \epsilon \quad \text{if} \quad 0 < x - b < \delta \quad \text{and} \quad \left|\frac{1}{x} - \frac{1}{b}\right| < \epsilon \quad \text{if} \quad x \in \Delta.$$

The generalized limit definition applies to $\lim\limits_{b^+}$ (right-side limit at b) to yield a definition of $\lim\limits_{b^+} f(x) = L$ if $f = \{(x, (1/x))\}$, $x \neq 0$, and $L = 1/b$.

We could continue to investigate examples but it seems unnecessary to do so. These two examples and the exercises that follow should enable you to apply the generalized limit definition to each of the six types of particular limits.

At this point, the attitude of a reader might be, "Why discuss the generalized limit—six different limit definitions are enough to remember." The advantages of introducing the generalized limit definition are primarily pedagogical and a matter of convenience. In order to become better acquainted with the heart of the limit concept we have expressed one statement that encompasses each of the six limit definitions; this is pedagogically advantageous and also convenient. Even more convenient is the opportunity to express one "generalized limit theorem" that applies to six particular types of limits. For instance, the six theorems listed in the beginning of this section could be expressed in the single "generalized limit theorem": If *limit* $f(x) = L$ and *limit* $f(x) = M$, then $L = M$. If we could "prove" one "generalized limit theorem," we could be, in effect, proving six particular limit theorems. We will seek ways of assuring these six-for-one returns in the next section. However, to prepare for an extension of the generalized limit concept let us return to the problem at hand: becoming better acquainted with the generalized limit definition.

Exercises

For Exercises 1 *and* 2 *consider* $f = \{(x, (1/x))\}$, $x \neq 0$, *and* $\lim_{\rightarrow} (1/x) = 0$.

1. If $\epsilon = \frac{1}{10}$ there is a δ-set, call it Δ_1, such that $|(1/x) - 0| < \frac{1}{10}$ if $x \in \Delta_1$. Give *three* different examples of Δ_1.

2. If $\epsilon = \frac{1}{100}$ there is a δ-set, call it Δ_2, such that $|(1/x) - 0| < \frac{1}{100}$ if $x \in \Delta_2$. Give *three* different examples of Δ_2.

3. Take any pair of δ-sets Δ_1 and Δ_2 named in Exercises 1 and 2. Indicate the set of real numbers x that are in both Δ_1 and Δ_2.

4. Let the generalized limit definition be applied to the limit at b of a function f. Give examples of the δ-sets that apply to the generalized limit definition when **(a)** $b = 2$; **(b)** $b = 3$; **(c)** $b = 0$.

For Exercises 5 *through* 10 *prove that the given domain* D_f *contains at least one δ-set for each of these parts of Figure* 3–2. (*Sample solutions are provided for Exercises* 5 *and* 8.)

5. For I, $\lim$, D_f is the set of natural numbers.

   Solution: D_f contains the set of natural numbers in $(1, \rightarrow)$ and in general the set of natural numbers in $(n, \rightarrow)$ for every natural number n; each of these sets is a δ-set accociated with $\lim$. Hence, D_f contains at least one δ-set.

6. For II, $\lim_{\rightarrow}$, D_f contains an open right ray $(p, \rightarrow)$.

7. For III, $\lim_{\leftarrow}$, D_f contains an open left ray $(\leftarrow, q)$.

8. For IV, $\lim_{b^+}$, D_f contains an open interval (b, c).

   Solution: D_f contains $\left(b, \frac{1}{2}(b + c)\right)$ and in general $(b, b + \delta)$ where $0 < \delta < c - b$; each of these sets is a δ-set associated with $\lim_{b^+}$. Hence, D_f contains at least one δ-set.

9. For V, $\lim_{b^-}$, D_f contains an open interval (a, b).

10. For VI, $\lim_{b}$, D_f contains a deleted neighborhood of b, such as

$$[(a, b) \cup (b, c)].$$

For Exercises 11 *through* 16 *prove that the given δ-set* Δ *is a non-empty set for each of these parts of Figure* 3–2. (*Sample solutions are provided for Exercises* 11 *and* 14.)

11. For I, $\lim$, Δ is the set of natural numbers in $(\delta, \rightarrow)$.

    Solution: For every real number $\delta > 0$ there exists some natural number n such that $n > \delta$. Hence, $n \in (\delta, \rightarrow)$; thus, Δ is non-empty.

12. For II, $\lim_{\rightarrow}$, Δ is the set of real numbers in $(\delta, \rightarrow)$.

13. For III, $\underset{\leftarrow}{lim}$, Δ is the set of real numbers in $(\leftarrow, -\delta)$.

14. For IV, $\underset{b^+}{lim}$, Δ is the set of real numbers in $(b, b + \delta)$.

Solution: Since $\delta > 0$, then $\delta > \frac{1}{2}\delta > 0$. Hence, $b + \frac{1}{2}\delta \in (b, b + \delta)$; thus, Δ is non-empty.

15. For V, $\underset{b^-}{lim}$, Δ is the set of real numbers in $(b - \delta, b)$.

16. For VI, $\underset{b}{lim}$, Δ is the set of real numbers in $[(b - \delta, b) \cup (b, b + \delta)]$.

3-3 The Generalized Limit

In §3–2 we considered all six of the different limit definitions encountered in Chapters 1 and 2. We compared the various definitions in Figure 3–1 and noted similarities. We listed some details (L is a real number, f is a function, etc.) that were to be assumed in our general discussion but verified in application. Then, in order to spotlight the similarities of the limit definitions and conveniently refer to the six different types of limits, we presented the generalized limit definition:

> ***Limit** f(x) = L* **iff for every** $\epsilon > 0$ **there exists a** δ**-set** Δ **such that**
>
> $$|f(x) - L| < \epsilon \text{ if } x \in \Delta.$$

In this section we intend to apply our generalized limit concept to include theorems. How will we do this? You, as a reader, may have some thoughts of your own. If so, you should try them out for yourself before reading further, and you may avoid spoiling your chance to be original. To provide food for thought, let us consider the six theorems concerning the uniqueness of particular limits. The theorems, listed in §3–2, are all similar in form to the single statement:

> *If limit f(x) = L and limit f(x) = M, then L = M.*

We could replace "*limit*" with "$\underset{\rightarrow}{lim}$" in this statement and have Theorem 2–8a; replace "*limit*" with "$\underset{\leftarrow}{lim}$" and have Theorem 2–8b; replace "*limit*" with "$\underset{b}{lim}$" and have Theorem 2–12c; and so forth. This single statement encompasses all six of the theorems concerning particular limits.

Could a proof be devised that would encompass all six of the proofs of the six individual theorems? Let us consider a particular proof, for instance the proof of Theorem 1–11a, and translate it into the language of the generalized limit definition:

"**Theorem**" *If limit f(x) = L and limit f(x) = M, then L = M.*

"*Proof:*" We shall use the indirect method of proof. Assume $L \neq M$. Let $L > M$, then $L - M > 0$. Since *limit f(x) = L*, the generalized limit definition implies the following: For every $\epsilon > 0$, and in particular $\epsilon = \frac{1}{2}(L - M)$, there exists a δ-set Δ_1 such that if $x \in \Delta_1$, then $|f(x) - L| < \frac{1}{2}(L - M)$ and

$$L - \tfrac{1}{2}(L - M) < f(x) < L + \tfrac{1}{2}(L - M). \tag{1}$$

Since *limit f(x) = M*, the generalized limit definition implies the following: For every $\epsilon > 0$, and in particular $\epsilon = \frac{1}{2}(L - M)$, there exists a δ-set Δ_2 such that if $x \in \Delta_2$, then $|f(x) - M| < \frac{1}{2}(L - M)$ and

$$M - \tfrac{1}{2}(L - M) < f(x) < M + \tfrac{1}{2}(L - M). \tag{2}$$

Consider the set Δ_3 of numbers x such that $x \in \Delta_1$ and $x \in \Delta_2$; that is, consider Δ_3 where $\Delta_3 = \Delta_1 \cap \Delta_2$. When $x \in \Delta_3$, both (1) and (2) will hold; and by (1) $\frac{1}{2}(L + M) = L - \frac{1}{2}(L - M) < f(x)$
and by (2) $f(x) < M + \frac{1}{2}(L - M) = \frac{1}{2}(L + M)$.
Therefore, when $x \in \Delta_3$, $\frac{1}{2}(L + M) < f(x) < \frac{1}{2}(L + M)$. But this inequality is an absurdity because $\frac{1}{2}(L + M) = \frac{1}{2}(L + M)$. Therefore, the assumption $L \neq M$ is impossible, and we must have $L = M$.

The "Theorem" and "proof" above would apply to Theorem 1–11a for uniqueness of a limit of a sequence if instead of "*limit*" we wrote "*lim*" and instead of "$x \in \Delta_1$" we wrote "$x > \delta_1$", where x is an integer and δ_1 is a positive real number (and the same for δ_2 and δ_3). The δ-set Δ_3 would be the set of natural numbers x such that $x > \delta_3$ where $\delta_3 = \max [\delta_1, \delta_2]$. A review of the proof of Theorem 2–8a for *lim* reveals that our "proof" would apply to $\underset{\rightarrow}{lim}$ in a similar manner. Furthermore, what if "$x \in \Delta_1$" were replaced with "$0 < |x - b| < \delta_1$" (and the same for Δ_2 and Δ_3)? Would the "proof" apply to Theorem 2–12c for $\underset{b}{lim}$? For Theorem 2–12c, however, the δ-set Δ_3 would be the set of real numbers x such that $0 < |x - b| < \delta_3$ where $\delta_3 = \min [\delta_1, \delta_2]$. The answer to the question is "yes", as would be the answer to each of six questions concerning the application of this "proof" (or "general argument" if you wish to call it that) to the six theorems concerning particular limits.

The applicability of our general "Theorem" to each of the six theorems for particular limits is dependent upon the properties of δ-sets. In the "proof" we selected a δ-set Δ_3 that enabled us to complete our argument. It was necessary that Δ_3 be a non-empty δ-set of numbers x such that if $x \in \Delta_3$, then $x \in \Delta_1$ and $x \in \Delta_2$. If you tested the "Theorem" for each type of limit you would find that in each case a properly chosen set Δ_3 possesses these properties.

Let us investigate δ-sets and determine in general that the δ-sets associated with any particular limit of a function satisfy the following **properties of δ-sets.** If a function f has a limit ($\underset{\rightarrow}{lim}$, $\underset{\leftarrow}{lim}$, $\underset{b^+}{lim}$, $\underset{b^-}{lim}$, $\underset{b}{lim}$, or lim), then there exist δ-sets associated with the limit that are subsets of the domain of f. This statement was verified in Exercises 5 through 10 of §3–2. Thus we state

Property 1 of δ-sets *The domain of any function that has a limit (any particular limit of the six types we have discussed) contains at least one δ-set associated with that limit. Thus, the set of δ-sets is non-empty.*

In Exercises 11 through 16 of §3–2 we proved that every δ-set was non-empty. Hence, we state

Property 2 of δ-sets *Every δ-set associated with each limit (any particular limit of the six types discussed) is non-empty. Thus, every δ-set is non-empty.*

The final property, the one that is of greatest importance to us, has been discussed but not verified. In our preceding "proof," Δ_3 was defined to be the set of real numbers x such that $x \in \Delta_1$ and $x \in \Delta_2$; that is, $\Delta_3 = \Delta_1 \cap \Delta_2$. Therefore, since every element of Δ_3 is an element of Δ_1, we say that Δ_3 is a *subset* of Δ_1 and we write $\Delta_3 \subset \Delta_1$, and similarly $\Delta_3 \subset \Delta_2$. However, we assumed that Δ_3 is a δ-set, and we should see to it that this is indeed true. Then we could assert

Property 3 of δ-sets *If Δ_1 and Δ_2 are two δ-sets associated with the same particular limit and $\Delta_3 = \Delta_1 \cap \Delta_2$, then Δ_3 is a δ-set for the same limit, and furthermore $\Delta_3 \subset \Delta_1$ and $\Delta_3 \subset \Delta_2$. Thus, $\Delta_1 \cap \Delta_2$ is a δ-set.*

From the definition of the intersection of two sets it follows that if $\Delta_3 = \Delta_1 \cap \Delta_2$, then $\Delta_3 \subset \Delta_1$ and $\Delta_3 \subset \Delta_2$. Hence, to verify Property 3 of δ-sets we need only to show that Δ_3 is a δ-set for each type of limit. For limit types I and II, lim and $\underset{\rightarrow}{lim}$, let δ_1 and δ_2 be positive real numbers, and let Δ_1 and Δ_2 be δ-sets $(\delta_1, \rightarrow)$ and $(\delta_2, \rightarrow)$, respectively. Then

$$\Delta_3 = \Delta_1 \cap \Delta_2 = \{x | x > \delta_3 \quad \text{where} \quad \delta_3 = \max [\delta_1, \delta_2]\}$$

and hence, the set of natural numbers in $(\delta_3, \rightarrow)$ is a δ-set for lim and the set of real numbers in $(\delta_3, \rightarrow)$ is a δ-set for $\underset{\rightarrow}{lim}$.

Similarly, for limit type IV, $\underset{b^+}{lim}$, let Δ_1 and Δ_2 be the δ-sets $(b, b + \delta_1)$ and $(b, b + \delta_2)$, respectively. Then

$$\Delta_3 = \Delta_1 \cap \Delta_2 = \{x | 0 < x - b < \delta_3 \quad \text{where} \quad \delta_3 = \min [\delta_1, \delta_2]\},$$

and hence, Δ_3 is the δ-set $(b, b + \delta_3)$ for $\underset{b^+}{lim}$. The verification of Property 3 of δ-sets is completed in Exercise 1 where $\underset{\leftarrow}{lim}$, $\underset{b^-}{lim}$, and $\underset{b^+}{lim}$ are considered.

Now that we have described δ-sets, think again of the statement: If *limit f(x) = L* and *limit f(x) = M*, then $L = M$. The "proof" that we gave involved the generalized limit definition and Properties 1, 2, and 3 of δ-sets. The generalized limit definition applies to each particular limit of the six types we defined, and Properties 1, 2, and 3 of δ-sets are verified for each particular limit. Therefore, our so called "proof" is indeed a proof for each one of the six individual theorems; the real difference between our "proof" and the proof of any one of the theorems is the notation we use (for example, for *lim* we use "*limit*" instead of "*lim*," "δ-set" instead of "$x > δ$," etc.).

If a statement involves the generalized limit definition and can be proved with Properties 1, 2, and 3 of δ-sets, we shall call it a **generalized limit theorem**. As we have remarked above, the proof of a generalized limit theorem provides a proof for six theorems concerning particular limits. Our first generalized limit theorem is therefore,

Theorem 3-3a *If limit f(x) = L, and limit f(x) = M, then L = M.*

Some other generalized limit theorems are the following:

Theorem 3-3b *If limit f(x) = L and f(x) = g(x) for each x in some δ-set, then limit g(x) = L.*

Proof: (Of course, we must assume that *f* and *g* are functions, *L* is a real number, and, for any particular limit, that the domains of *f* and *g* are appropriate; but we had previously agreed that these details would be assumed.) Let any $\epsilon > 0$ be given. We intend to find a δ-set Δ such that

$$|g(x) - L| < \epsilon \quad \text{if} \quad x \in \Delta.$$

Since *limit f(x) = L*, by definition there exists a δ-set Δ_1 such that

$$|f(x) - L| < \epsilon \quad \text{if} \quad x \in \Delta_1.$$

Since we are given that $f(x) = g(x)$ for all *x* in some δ-set, let us say

$$f(x) = g(x) \quad \text{if} \quad x \in \Delta_2.$$

Let $\Delta_3 = \Delta_1 \cap \Delta_2$. By the properties of δ-sets Δ_3 is a δ-set; therefore, both $|f(x) - L| < \epsilon$ and $f(x) = g(x)$ if $x \in \Delta_3$. It follows then that

$$|g(x) - L| < \epsilon \quad \text{if} \quad x \in \Delta_3.$$

Thus, for every $\epsilon > 0$ there exists a δ-set Δ (determined as Δ_3) such that

$$|g(x) - L| < \epsilon \quad \text{if} \quad x \in \Delta;$$

that is, *limit g(x) = L*.

Example 1 Let $f = \{(x, x)\}$ and $g = \{(x, |x|)\}$. Note that $\lim_{-2} f(x) = -2$. Show that Theorem 3–3b can or cannot be used to imply that $\lim_{-2} g(x) = -2$.

Solution: $\underset{-2}{Lim}\, f(x) = -2$, and $f(x) = g(x)$ if $x \geq 0$. However, $f(x) \neq g(x)$ on any deleted neighborhood of -2. Since deleted neighborhoods of -2 are δ-sets for $\underset{-2}{lim}$, the hypothesis of Theorem 3–3b is not met and the theorem does not apply. Note also that $\underset{-2}{lim}\, g(x) = 2 \neq -2$

Theorem 3-3c *If* $limit\, f(x) = L$, *then* f *is defined and bounded on some* δ-*set.*

Proof: Let $\epsilon = \frac{1}{2}$. We are given that $limit\, f(x) = L$. Hence, by the definition of limit we know there exists some δ-set Δ such that f is defined on Δ and

$$|f(x) - L| < \tfrac{1}{2} \quad \text{if} \quad x \in \Delta.$$

But, $|f(x) - L| < \frac{1}{2}$ is equivalent to $L - \frac{1}{2} < f(x) < L + \frac{1}{2}$. Hence,

$$L - \tfrac{1}{2} < f(x) < L + \tfrac{1}{2} \quad \text{if} \quad x \in \Delta.$$

Thus, we have determined that there is a δ-set Δ such that f is defined on Δ and f is bounded (below by $L - \frac{1}{2}$, above by $L + \frac{1}{2}$) on Δ.

Example 2 Show how Theorem 3–3c and its proof apply to $\underset{\rightarrow}{lim}\, 1/x$.

Solution: Since $\underset{\rightarrow}{lim}\, (1/x) = 0$, the hypothesis of Theorem 3–3c is satisfied. Let $\epsilon = \frac{1}{2}$. Then there exists a δ-set $\Delta = (2, \rightarrow)$ such that $(1/x)$ is a real number for every $x \in \Delta$ and

$$\left|\frac{1}{x} - 0\right| < \frac{1}{2} \quad \text{if} \quad x \in \Delta;$$

that is, if $x > 2$, then $(1/x)$ is a real number and $|(1/x) - 0| < \frac{1}{2}$. Hence, if $x \in \Delta \ (x > 2)$, then

$$-\frac{1}{2} < \frac{1}{x} < \frac{1}{2}.$$

Therefore, f is defined and bounded (below by $-\frac{1}{2}$, above by $\frac{1}{2}$) on Δ where $\Delta = (2, \rightarrow)$.

The following exercises contain generalized limit theorems. For particular cases, similar theorems were proved in Chapters 1 or 2. Upon completing this section we should be somewhat appreciative of the generalized limit concept if we realize that Theorems 3–3a, 3–3b, and 3–3c, along with the six theorems in the exercises, actually provide us with fifty-four $[6(3 + 6) = 54]$ limit theorems for particular limits.

Exercises

1. Verify Property 3 of δ-sets for: (a) $\lim_{\leftarrow}$; (b) $\lim_{b^-}$; (c) $\lim_{b}$.

For Exercises 2 through 7, prove the following generalized limit theorems.

2. If $f = \{(x, c)\}$ (f is a constant function), then $limit\ c = c$.

3. If $limit\ g(x) = p > 0$, there exists some δ-set Δ such that $g(x) > \frac{1}{2}p > 0$ for every $x \in \Delta$.

4. If $limit\ g(x) = -p < 0$, there exists some δ-set Δ such that $g(x) < -\frac{1}{2}p < 0$ for every $x \in \Delta$.

5. If $limit\ g(x) = M \neq 0$, then there exists a δ-set Δ such that $|g(x)| > \frac{1}{2}|M|$ if $x \in \Delta$. *Hint:* If $M > 0$, let $M = p > 0$ and use Exercise 3. If $M < 0$, let $M = -p < 0$ and use Exercise 4.

6. *Limit* $f(x) = L$ iff $limit\ (f(x) - L) = 0$.

7. Let $limit\ f(x) = limit\ h(x) = L$ and $f(x) \leq g(x) \leq h(x)$ for every x in some δ-set Δ. Prove that $limit\ g(x) = L$. *Hint:* For any $\epsilon > 0$, you can obtain $L - \epsilon < f(x)$ and $h(x) < L + \epsilon$ and $f(x) \leq g(x) \leq h(x)$ for every x in some δ-set Δ. (This is a general statement of the **domination principle**.)

3-4 Generalized Limit Theorems

In this chapter we have planted two conceptual seeds: the arithmetic of functions and the generalized limit. Through study of definitions, examples, and exercises these concepts should have grown and matured in your mind. Now it is time to bring in the harvest.

We can "break-down" some functions as the sum, difference, product, or quotient of more simple functions. If we knew the limits of the simpler functions, could we determine the limit of their sum, difference, product, or quotient? For example,

If $\lim \dfrac{n-1}{n} = 1$, and $\lim \dfrac{1}{n} = 0$; what is $\lim \left(\dfrac{n-1}{n} + \dfrac{1}{n} \right)$?

If $\lim_{\leftarrow} \dfrac{1}{x} = 0$, and $\lim_{\leftarrow} \dfrac{x-1}{x} = 1$; what is $\lim_{\leftarrow} \left(\dfrac{1}{x} - \dfrac{x-1}{x} \right)$?

If $\lim_{\rightarrow} 2 = 2$, and $\lim_{\rightarrow} \dfrac{1}{x} = 0$; what is, $\lim_{\rightarrow} \left(2 \times \dfrac{1}{x} \right)$?

If $\lim_{4} \sqrt{x} = 2$, and $\lim_{4} x = 4$; what is $\lim_{4} \left(\dfrac{\sqrt{x}}{x} \right)$?

Each of the questions and infinitely many more could be answered individually by finding a number, calling it L, and verifying that the particular limit definition holds for L. The most efficient approach for us, however, would be to utilize our generalized limit concept. Then, by proving one theorem for the generalized limit of an arithmetic combination of two functions we would gain six theorems, each of which would apply to six types of limits.

Our task is outlined, but before undertaking the proofs of the generalized limits one detail must be stressed. Undoubtedly, we have established some ideas concerning such examples as:

If $\lim_{4} f(x) = L$, and $\lim_{4} g(x) = M$, then $\lim_{4} \big(f(x) + g(x)\big) = L + M$;

or, if $\lim_{4} f(x) = L$, and $\lim_{4} g(x) = M \neq 0$, then $\lim_{4} \big(f(x) \div g(x)\big) = L \div M$.

However, we cannot draw such an obvious conclusion from the joint assumption of the following pairs of limits:

$$\lim_{4} f(x) = L, \qquad \lim_{\rightarrow} g(x) = M;$$
$$\lim_{4} f(x) = L, \qquad \lim_{5} g(x) = M;$$
$$\lim_{\rightarrow} f(x) = L, \qquad \lim_{\leftarrow} g(x) = M.$$

Therefore, let us make the following agreement that will enable us to apply our generalized limit concept to theorems involving two functions:

Any generalized limit theorem that involves two functions pertains only to one particular limit unless specified otherwise.

Now let us seek to establish some theorems concerning arithmetic combinations of functions. Preceding the proof of some of the following theorems we have included some informal discussion and some questions. You, the reader, may be able to prove a theorem by yourself after you have thought through the discussion, or at least the proof may seem more natural after you have been exposed to selected questions.

Theorem 3-4a If $limit\, f(x) = L$ and $limit\, g(x) = M$, then $limit\, \big(f(x) + g(x)\big) = L + M$.

Discussion: The definition of $limit\, \big(f(x) + g(x)\big) = L + M$ involves the inequality

$$\big|\big(f(x) + g(x)\big) - (L + M)\big| < \epsilon.$$

How is $\big|\big(f(x) + g(x)\big) - (L + M)\big|$ related to $|f(x) - L|$ and $|g(x) - M|$?

If $|f(x) - L| < \frac{1}{2}\epsilon$ and $|g(x) - M| < \frac{1}{2}\epsilon$, what could you say about $|(f(x) + g(x)) - (L + M)|$?

Proof: Let any $\epsilon > 0$ be given. We must determine some δ-set Δ such that

$$|(f(x) + g(x)) - (L + M)| < \epsilon \qquad \text{if} \qquad x \in \Delta.$$

Since $limit\, f(x) = L$, $limit\, g(x) = M$, and $\frac{1}{2}\epsilon > 0$, the generalized limit definition implies the following: There exists a δ-set Δ_1 such that if $x \in \Delta_1$, then

$$|f(x) - L| < \tfrac{1}{2}\epsilon; \tag{1}$$

and there exists a δ-set Δ_2 such that if $x \in \Delta_2$, then

$$|g(x) - M| < \tfrac{1}{2}\epsilon. \tag{2}$$

From the properties of a δ-set we are assured that $\Delta_1 \cap \Delta_2$ is a non-empty δ-set that is a subset of both Δ_1 and Δ_2. Hence, if $\Delta_3 = \Delta_1 \cap \Delta_2$ and $x \in \Delta_3$, then both (1) and (2) hold to yield

$$|f(x) - L| + |g(x) - M| < \tfrac{1}{2}\epsilon + \tfrac{1}{2}\epsilon = \epsilon.$$

But $|(f(x) + g(x)) - (L + M)| \leq |f(x) - L| + |g(x) - M|$ and, thus,

$$|(f(x) + g(x)) - (L + M)| < \epsilon \text{ if } x \in \Delta_3.$$

Thus, Δ_3 is the δ-set we were to determine. Therefore, we can conclude: For every $\epsilon > 0$ there exists a δ-set Δ (determined as Δ_3) such that

$$|(f(x) + g(x)) - (L + M)| < \epsilon \qquad \text{if} \qquad x \in \Delta;$$

that is, $limit\, (f(x) + g(x)) = L + M$.

Example 1 Rephrase the proof of Theorem 3–4a for the theorem: If $\lim_{\rightarrow} f(x) = L$ and $\lim_{\rightarrow} g(x) = M$, then $\lim_{\rightarrow} (f(x) + g(x)) = L + M$.

Solution: To obtain a proof for this particular theorem make the following changes in the proof of Theorem 3–4a: for "δ-set Δ," write "real number $\delta > 0$"; for "$x \in \Delta$," write "$x > \delta$" (do the same for Δ_1, Δ_2, and Δ_3); for the discussion concerning Δ_3, write "choose $\delta_3 = \max [\delta_1, \delta_2]$." Then Theorem 3–4a and its proof will be in the language of the definition of $\lim_{\rightarrow}$.

Example 2 Determine $\lim_{0} (\sin x + \cos x)$.

Solution: We have shown that $\lim_{0} \sin x = 0$ and $\lim_{0} \cos x = 1$ (see §2–13). Hence, by Theorem 3–4a $\lim_{0} (\sin x + \cos x) = 0 + 1 = 1$.

Theorem 3-4b *If* $limit\, f(x) = L$ *and* $limit\, g(x) = M$, *then* $limit\, (f(x) - g(x)) = L - M$.

Discussion: In the real number system, $L - M = L + (-M)$. Hence, if we first prove $limit\,(-g(x)) = -M$, we can use Theorem 3–4a to obtain:

$$limit\,(f(x) - g(x)) = limit\,(f(x) + [-g(x)]) = L + (-M) = L - M.$$

We shall prove $limit\,(-g(x)) = -M$ whenever $limit\,g(x) = M$ as a special case of Theorem 3–4c. This will complete the proof of Theorem 3–4b. A proof that is analogous to that of Theorem 3–4a could be given, but this is an excellent exercise (see Exercise 1).

Theorem 3-4c *If limit $g(x) = M$, then limit $kg(x) = kM$ for any real number $k \neq 0$.*

Discussion: From the generalized limit definition we gain information concerning $|g(x) - M|$. We seek information concerning $|kg(x) - kM|$; that is, about $|k||g(x) - M|$. Under what conditions on $|g(x) - M|$ is

$$|k||g(x) - M| < \epsilon?$$

Proof: Let any $\epsilon > 0$ be given. We must determine some δ-set Δ such that

$$|kg(x) - kM| < \epsilon \quad \text{if} \quad x \in \Delta.$$

We have $limit\,g(x) = M$ and $(\epsilon/|k|) > 0$. Therefore there is a δ-set Δ_1 such that if $x \in \Delta_1$, then

$$|g(x) - M| < \frac{\epsilon}{|k|},$$

$$|k||g(x) - M| < \epsilon,$$

and $\qquad\qquad |kg(x) - kM| < \epsilon.$

Therefore, we can conclude: For every $\epsilon > 0$ there exists a δ-set Δ (determined as Δ_1) such that

$$|kg(x) - kM| < \epsilon \quad \text{if} \quad x \in \Delta;$$

that is, $limit\,kg(x) = kM$.

Theorem 3-4d *If limit $f(x) = L$ and limit $g(x) = M$, then limit $(f(x) \times g(x)) = L \times M$.*

Discussion: From the generalized limit definitions we gain information concerning $|f(x) - L|$ and $|g(x) - M|$. For the limit definition we seek information concerning $|(f(x) \times g(x)) - (L \times M)|$ which is conveniently written $|f(x)g(x) - LM|$. However,

$$(f(x) - L)(g(x) - M) = f(x)\,g(x) - Lg(x) - Mf(x) + LM. \qquad (3)$$

We want $|f(x)\,g(x) - LM|$ to be related to $|f(x) - L|$ and $|g(x) - M|$, but there is no apparent relation in (3).

A novice "theorem prover" might not realize how to proceed. One of a variety of possibilities is to add $0 = -Lg(x) + Lg(x)$, so that

$$\begin{aligned}|f(x)g(x) - LM| &= |f(x)g(x) - Lg(x) + Lg(x) - LM| \\ &= |(f(x) - L)g(x) + L(g(x) - M)| \\ &\leq |f(x) - L||g(x)| + |L||g(x) - M|.\end{aligned}$$

Now, $|L||g(x) - M| < \epsilon$ under what conditions on $|g(x) - M|$? If $|g(x)| < B$, what conditions imposed on $|f(x) - L|$ will insure that $|f(x) - L||g(x)| < \epsilon$?

Proof: Let any $\epsilon > 0$ be given. We must determine a δ-set Δ such that

$$|f(x)g(x) - LM| < \epsilon \quad \text{if} \quad x \in \Delta.$$

Since
$$\begin{aligned}|f(x)g(x) - LM| &= |f(x)g(x) - Lg(x) + Lg(x) - LM| \\ &= |(f(x) - L)g(x) + L(g(x) - M)| \\ &\leq |f(x) - L||g(x)| + |L||g(x) - M|,\end{aligned}$$

a δ-set Δ will suffice if for every $x \in \Delta$

$$|f(x) - L||g(x)| < \tfrac{1}{2}\epsilon \tag{4}$$

and
$$|L||g(x) - M| < \tfrac{1}{2}\epsilon. \tag{5}$$

Since *limit* $g(x) = M$, g is defined and bounded on some δ-set by Theorem 3-3c; that is, there is some δ-set Δ_1 and some real number $B > 0$ such that if $x \in \Delta_1$, then

$$|g(x)| < B. \tag{6}$$

Since *limit* $f(x) = L$ and $(\epsilon/2B) > 0$, there exists some δ-set Δ_2 such that if $x \in \Delta_2$, then

$$|f(x) - L| < \frac{\epsilon}{2B}. \tag{7}$$

Select $\Delta_3 = \Delta_1 \cap \Delta_2$, and both (6) and (7) are true for $x \in \Delta_3$. Hence, by multiplication, if $x \in \Delta_3$ we obtain (4):

$$|f(x) - L||g(x)| < \frac{\epsilon}{2}.$$

If $L = 0$, there is a δ-set Δ_4 (for instance, $\Delta_4 = \Delta_3$) such that equation (5) is certainly true. If $L \neq 0$ we proceed as follows: Since *limit* $g(x) = M$ and $(\epsilon/2|L|) > 0$, there exists another δ-set Δ_4 such that if $x \in \Delta_4$, then

$$|g(x) - M| < \frac{\epsilon}{2|L|}.$$

Hence, if $x \in \Delta_4$, we multiply by $|L|$ and obtain (5):

$$|L||g(x) - M| < \frac{\epsilon}{2}.$$

By the properties for δ-sets choose $\Delta_5 = \Delta_3 \cap \Delta_4$, and both (4) and (5) are true if $x \in \Delta_5$. Hence, by adding (4) and (5) we obtain

$$|f(x) - L||g(x)| + |L||g(x) - M| < \epsilon.$$

But, $|f(x) g(x) - LM| \leq |f(x) - L||g(x)| + |L||g(x) - M|$, and it follows that

$$|f(x)g(x) - LM| < \epsilon \qquad \text{if} \qquad x \in \Delta_5.$$

Therefore, for every $\epsilon > 0$ there exists a δ-set Δ (determined as Δ_5) such that

$$|f(x)g(x) - LM| < \epsilon \qquad \text{if} \qquad x \in \Delta;$$

that is, $limit\, f(x)\, g(x) = LM$.

Example 3 Use Theorem 3–4d to evaluate $\lim_b x^n$ for any natural number n.

> *Solution:* We have previously shown (Exercise 6 of §2–12) that $\lim_b x = b$. Hence, by repeated use of Theorem 3–4d, $\lim_b x^2 = \lim_b (x \cdot x) = b \cdot b = b^2$; $\lim_b x^3 = \lim_b (x \cdot x^2) = b \cdot b^2 = b^3$; and if $\lim_b x^k = b^k$, then $\lim_b x^{k+1} = \lim_b (x \cdot x^k) = b \cdot b^k = b^{k+1}$. Hence, $\lim_b x^n = b^n$ for any natural number n.

Theorem 3-4e *If $limit\, f(x) = L$ and $limit\, g(x) = M \neq 0$, then $limit\, (f(x) \div g(x)) = L \div M$.*

Discussion: In the real number system $L \div M = L(1/M)$, $M \neq 0$. Our approach to this theorem parallels the approach to the "subtraction theorem" Theorem 3–4b. If we can prove $limit\, (1/g(x)) = 1/M$ (given $limit\, g(x) = M \neq 0$ and $limit\, f(x) = L$) we could use Theorem 3–4d to obtain:

$$limit\, (f(x) \div g(x)) = limit\, \left(f(x) \times \frac{1}{g(x)}\right) = L\left(\frac{1}{M}\right) = L \div M.$$

We will complete this proof by verifying in Theorem 3–4f that $limit\, (1/g(x)) = 1/M$ whenever $limit\, g(x) = M \neq 0$.

Theorem 3-4f *If $limit\, g(x) = M \neq 0$, then $limit\, \dfrac{1}{g(x)} = \dfrac{1}{M}$.*

Discussion: The definition of $limit\, g(x) = M$ gives us information concerning $|g(x) - M|$. We seek information concerning $\left|\dfrac{1}{g(x)} - \dfrac{1}{M}\right|$. By simple addition of fractions

$$\left|\frac{1}{g(x)} - \frac{1}{M}\right| = \left|\frac{M - g(x)}{g(x)M}\right| = \frac{1}{|g(x)|}\frac{1}{|M|}|g(x) - M|.$$

Recall that we proved in Exercise 5 of §3–3 that if $limit\, g(x) = M \neq 0$, then $|g(x)| > \frac{1}{2}|M|$ for every x in some δ-set. Can a condition be placed on $|g(x) - M|$ to imply that $\dfrac{1}{|g(x)|}\dfrac{1}{|M|}|g(x) - M| < \epsilon$?

Proof: Let any $\epsilon > 0$ be given. We must determine a δ-set Δ such that

$$\left| \frac{1}{g(x)} - \frac{1}{M} \right| < \epsilon \qquad \text{if} \qquad x \in \Delta.$$

We are given *limit* $g(x) = M \neq 0$. Hence, by Exercise 5 of §3–3 there is some δ-set Δ_1 such that if $x \in \Delta_1$, then $|g(x)| > \frac{1}{2}|M|$

and
$$\frac{1}{|g(x)|} < \frac{2}{|M|}. \tag{8}$$

Also, $(|M|^2\epsilon/2) > 0$. Hence, by the limit definition there exists a δ-set Δ_2 such that if $x \in \Delta_2$, then

$$|g(x) - M| < \frac{|M|^2\epsilon}{2}. \tag{9}$$

Select $\Delta_3 = \Delta_1 \cap \Delta_2$, and both (8) and (9) are true; that is, if $x \in \Delta_3$, then, upon multiplying (8) and (9) and multiplying by $1/|M|$ we have

$$\frac{1}{|g(x)|}\frac{1}{|M|}|g(x) - M| < \frac{2}{|M|}\frac{1}{|M|}\frac{|M|^2\epsilon}{2} = \epsilon.$$

But $\left| \dfrac{1}{g(x)} - \dfrac{1}{M} \right| = \dfrac{1}{|g(x)|}\dfrac{1}{|M|}|g(x) - M|$, and it follows that

$$\left| \frac{1}{g(x)} - \frac{1}{M} \right| < \epsilon \qquad \text{if} \qquad x \in \Delta_3.$$

Therefore, for every $\epsilon > 0$ there exists a δ-set Δ (determined as Δ_3) such that

$$\left| \frac{1}{g(x)} - \frac{1}{M} \right| < \epsilon \qquad \text{if} \qquad x \in \Delta;$$

that is, *limit* $\dfrac{1}{g(x)} = \dfrac{1}{M}$.

Example 4 Evaluate $\lim\limits_{b} \dfrac{x^2 - b^2}{2x(x - b)}$, $b \neq 0$.

> *Solution:* To determine a limit at b we do not consider $x = b$. Hence,
> $\dfrac{x^2 - b^2}{2x(x - b)} = \dfrac{(x + b)(x - b)}{2x\,(x - b)} = \dfrac{x + b}{2x}$ because $x - b \neq 0$ for every value
> of x under consideration. Also, since $b \neq 0$, there is some deleted
> neighborhood of b in which $x \neq 0$. Since $\lim\limits_{b} x = b$, and $\lim\limits_{b} b = b$, then
> by Theorem 3–4a $\lim\limits_{b} (x + b) = 2b$. Furthermore, $\lim\limits_{b} 2x = 2b$ by Theorem
> 3–4c. Hence, by Theorem 3–4e, $\lim\limits_{b} \dfrac{x + b}{2x} = \dfrac{2b}{2b} = 1$ for $b \neq 0$; that is,
> $\lim\limits_{b} \dfrac{x^2 - b^2}{2x(x - b)} = 1$ for $b \neq 0$.

Exercises

For Exercises 1 *and* 2, *assume limit* $f(x) = L$ *and limit* $g(x) = M$.

1. Prove that $limit\,(f(x) - g(x)) = L - M$ by using the generalized limit definition.

2. Recall equation (3):

$$(f(x) - L)(g(x) - M) = f(x)g(x) - Lg(x) - Mf(x) + LM.$$

By Theorems 3–4a, 3–4b, and 3–4c,

$$limit\,(f(x)g(x) - Lg(x) - Mf(x) + LM) = limit\,f(x)g(x) - LM.$$

Prove $limit\,(f(x) - L)(g(x) - M) = 0$ from the generalized limit definition; then you will have an alternate proof of Theorem 3–4d.

3. Use the proved generalized limit theorems to determine the following limits. The first problem is worked as an example.

(a) $\displaystyle \lim_{b} \frac{\sqrt{x} + x^2}{x - 1}$, $b > 1$.

Solution: $\lim_{b} \sqrt{x} = \sqrt{b}$ (Example 1 of §2–12) and $\lim_{b} x^2 = b^2$ (Example 3 of §3–4); hence, $\lim_{b} (\sqrt{x} + x^2) = \sqrt{b} + b^2$ (Theorem 3–4a). Also, $\lim_{b} x = b$ (Exercise 6 of §2–12) and $\lim_{b} 1 = 1$; hence, $\lim_{b} (x - 1)$ $= b - 1$ (Theorem 3–4b). Therefore, $\displaystyle \lim_{b} \frac{\sqrt{x} + x^2}{x - 1} = \frac{\sqrt{b} + b^2}{b - 1}$ (Theorem 3–4e).

(b) $\displaystyle \lim_{b} (2x^3 - 3x + 7)$; 　　　　(c) $\displaystyle \lim_{b} \frac{x^2 - 1}{x^2 + 1}$;

(d) $\displaystyle \lim_{0} \frac{3 \sin^2 x}{x^2}$, Hint: Recall $\displaystyle \lim_{0} \frac{\sin x}{x} = 1$.

4. Prove that if $limit\,F(x) = 0$ and G is bounded on some δ-set (assume $B > 0$ and $|G(x)| < B$ if $x \in \Delta_1$), then $limit\,F(x)\,G(x) = 0$.

5. Use Exercise 4 to determine the following limits.

(a) $\displaystyle \lim_{0} \left(x \sin \frac{1}{x}\right)$; 　　　　(b) $\displaystyle \lim_{\to} \frac{\sin x}{x}$;

(c) $\displaystyle \lim_{5} \left((x - 5) \cos \frac{1}{x}\right)$; 　　　　(d) $\displaystyle \lim_{\leftarrow} \frac{|x|\,2^x}{x}$.

3-5 More on Composition and Continuity

In §2–14 we discussed continuity and defined a function to be continuous at b iff

(i) $f(b)$ is defined,

(ii) $\lim_{b} f(x)$ exists, and

(iii) $f(b) = \lim_{b} f(x)$.

We have proved that both $f = \{(x, x)\}$ and $g = \{(x, c)\}$ are continuous at b for every real number b. In particular,

$$\lim_b x = b = f(b), \qquad \text{and} \qquad \lim_b c = c = g(b).$$

These two functions, the identity function and a constant function, can be used to generate other functions and prove they are continuous.

Example 1 Let A, B, C be real numbers. Show that $h = \{(x, Ax^2 + Bx + C)\}$ is continuous at b for every real number b.

Solution: Since $\lim_b x = b$, then by Theorem 3–4d,

$$\lim_b x^2 = \lim_b (x \cdot x) = b \cdot b = b^2.$$

By Theorem 3–4c, $\lim_b Ax^2 = Ab^2$ and $\lim_b Bx = Bb$.

Since $\lim_b C = C$, then by Theorem 3–4a we can conclude,

$$\lim_b (Ax^2 + Bx + C) = Ab^2 + Bb + C.$$

Since $Ab^2 + Bb + C$ is also $h(b)$, all three conditions enumerated in the definition of continuity of a function are met; h is continuous at b for every real number b.

Example 1 can be extended and generalized to include any given polynomial in x as we show by the next theorem.

Theorem 3-5a *Let a_1, a_2, ..., a_n be real numbers and $1, 2, 3, ..., n$ be natural numbers. Then $f = \{(x, y)\}$ where*

$$y = a_n x^n + a_{n-1} x^{n-1} + a_{n-2} x^{n-2} + \cdots + a_1 x + a_0$$

is a continuous function at b for every real number b.

Proof: Since $\lim_b x = b$, repeated use of Theorem 3–4d implies

$$\lim_b x^k = b^k \ (k = 1, 2, 3, ..., n)$$

for each natural number k (see Example 3 of §3–4). By Theorem 3–4c,

$$\lim_b a_k x^k = a_k b^k \ (k = 1, 2, 3, ..., n).$$

Also, $\lim_b a_0 = a_0$. Now, repeated use of Theorem 3–4a yields

$$\lim_b (a_n x^n + a_{n-1} x^{n-1} + a_{n-2} x^{n-2} + \cdots + a_1 x + a_0) = f(b)$$

where $\qquad f(b) = a_n b^n + a_{n-1} b^{n-1} + a_{n-2} b^{n-2} + \cdots + a_1 b + a_0.$

Therefore, the polynomial function f is continuous at b where b is any real number.

Furthermore, the theorems on limits of arithmetic combinations of functions have analogies for continuity of functions.

Theorem 3-5b *If f and g are two functions continuous at b, then* (a) $f + g$, (b) $f - g$, (c) $f \times g$, (d) $f \div g$ *are continuous at b provided in* (d) *that* $\lim_{b} g(x) \neq 0$.

Proof: (a) By definition of continuity at b, $\lim_{b} f(x) = f(b)$, and $\lim_{b} g(x) = g(b)$ where $f(b)$ and $g(b)$ are real numbers. Thus, $f(b) + g(b)$ is a real number, and by Theorem 3–4a

$$\lim_{b} \big(f(x) + g(x) \big) = f(b) + g(b).$$

Hence, $f + g$ is continuous at b. (b), (c), (d) The proofs for these parts are analogous to the proof of part (a) except for their dependence on Theorem 3–4b, 3–4d, and 3–4e, respectively.

Example 2 Let $f = \{(x, x)\}$, and $g = \{(x, 5)\}$. Generate a function that is not a polynomial but that is continuous at every real number.

> *Solution:* One of many possible solutions is the following: Since f and g are continuous at every real number, then at every real number both $f \times f = \{(x, x^2)\}$ and $\big((f \times f) + g\big) = \{(x, x^2 + 5)\}$ are continuous. Also, $x^2 + 5 > 0$ for every real number. Hence, $g \div \big((f \times f) + g\big) = \left\{\left(x, \dfrac{5}{x^2 + 5}\right)\right\}$ is continuous at every real number, and this function is not a polynomial.

By using Theorem 3–5b, we can generate, as in Example 2, a large selection of continuous functions. However, since this task provides excellent experience, it is delegated to the exercises.

Let us consider another valuable extension of our understanding of continuity at b: an alternate definition. When we state the definition of $\lim_{b} f(x) = L$ we use the inequality

$$0 < |x - b| < \delta.$$

By discussing just those values of x for which $0 < |x - b|$ we make sure that b is not a considered value of x in the limit definition. However, for continuity at b we want the value b to be considered because, in the definition of continuity, (i) $f(b)$ must be defined and (iii) $\lim_{b} f(x) = f(b)$. Hence, the three conditions we specified in our limit definition are encompassed in the following alternate definition of continuity: A function f is **continuous at** b iff for every $\epsilon > 0$ there exists $\delta > 0$ such that

$$|f(x) - f(b)| < \epsilon \qquad \text{if} \qquad |x - b| < \delta.$$

If (i) $f(b)$ is defined, (ii) $\lim_{b} f(x)$ exists, and (iii) $\lim_{b} f(x) = f(b)$, then the alternate definition holds. Conversely, if $|f(x) - f(b)| < \epsilon$ when we consider $|x - b| < \delta$, then $|f(x) - f(b)| < \epsilon$ when we consider $0 < |x - b| < \delta$; thus the alternate definition implies (i) $f(b)$ exists since $|x - b| < \delta$ includes $x = b$, (ii) $\lim_{b} f(x)$ exists, and (iii) $\lim_{b} f(x) = f(b)$ by the limit definition.

Armed with this alternate definition we now undertake the other topic specified in our section title, composition. For an experience designed to permit you to activate your intuition and to stimulate your interest, consider Example 3.

Example 3 Determine $\lim_{\to} \cos (1/x)$ by analyzing the composite function $F \, of \, G$ where $G = \{(x, (1/x))\}$, $x \neq 0$ and $F = \{(z, \cos z)\}$. (We use z as the domain variable for F only to provide convenient reference to each of the two domains.)

Solution: Because we want the limit at-the-right, let us consider a set of increasing values of x; and determine $(1/x)$ and $\cos (1/x)$. Also, since we are finding $\cos (1/x)$, let $(1/x) = z$ and consider $\cos z = \cos (1/x)$.

x	$\dfrac{1}{\pi}$	$\dfrac{2}{\pi}$	$\dfrac{3}{\pi}$	$\dfrac{4}{\pi}$	$\dfrac{6}{\pi}$	$\dfrac{10}{\pi}$	$\dfrac{100}{\pi}$
$\dfrac{1}{x} = z$	π	$\dfrac{\pi}{2}$	$\dfrac{\pi}{3}$	$\dfrac{\pi}{4}$	$\dfrac{\pi}{6}$	$\dfrac{\pi}{10}$	$\dfrac{\pi}{100}$
$\cos z$	-1	0	$\dfrac{1}{2}$	$\dfrac{\sqrt{2}}{2}$	$\dfrac{\sqrt{3}}{2}$	$\approx .95106$	$\approx .99951$

Figure 3-4

We know that $\lim_{\to} (1/x) = 0$. Hence, $\lim_{\to} \cos (1/x)$ seems to depend upon $\lim_{0} \cos z$. Since $F = \{(z, \cos z)\}$ is continuous at 0, $\lim_{0} \cos z = 1$; and we propose that

$$\lim_{\to} \cos \frac{1}{x} = \lim_{0} \cos z = 1.$$

We shall prove that 1 is indeed the answer by establishing the next theorem.

In Example 3, we used the composite function $F \, of \, G$; that is, "cosine of reciprocal of x." Our answer then was dependent on

$$\lim_{\to} \frac{1}{x} = 0 \qquad \text{and} \qquad \lim_{0} \cos z = \cos 0 = 1.$$

Let us consider some other examples.

Example 4 Complete the sentences concerning $F = \{(z, \cos z)\}$, given functions g, and the composite functions $F \, of \, g$:

(a) Let $g = \{(x, 2^x)\}$. Since $\underset{\leftarrow}{lim}\ 2^x = 0$ and $\underset{0}{lim}\cos z = 1$,

then $\underset{\leftarrow}{lim}\cos 2^x = \boxed{\text{A}}$.

(b) Let $g = \{(x, x - 5)\}$. Since $\underset{5}{lim}(x - 5) = 0$ and $\underset{0}{lim}\cos z = 1$,

then $\boxed{\text{B}}\cos(x - 5) = 1$.

(c) Let $g = \left\{\left(x, \dfrac{|x|}{x} - 1\right)\right\}$, $x \neq 0$. Since $\underset{0^+}{lim}\left(\dfrac{|x|}{x} - 1\right) = 0$ and

$\underset{0}{lim}\cos z = 1$, then $\underset{0^+}{lim}\cos\boxed{\text{C}} = 1$.

> *Solution:* For A, use "1"; thus $\underset{\leftarrow}{lim}\cos 2^x = 1$. (b) For B, use "$\underset{5}{lim}$";
>
> thus $\underset{5}{lim}\cos(x - 5) = 1$. (c) For C, use "$\dfrac{|x|}{x} - 1$"; thus $\underset{0^+}{lim}\cos\left(\dfrac{|x|}{x} - 1\right) = 1$.

In general we might expect that $limit\ f(g(x)) = L$ would depend on

$$limit\ g(x) = b \qquad \text{and} \qquad \underset{b}{lim}\ f(z) = f(b) = L.$$

Indeed, let any $\epsilon > 0$ be given. By the alternate definition of continuity, $\underset{b}{lim}\ f(z) = f(b) = L$ implies there exists a real number $\delta > 0$ such that

$$|f(z) - L| < \epsilon \qquad \text{if} \qquad |z - b| < \delta.$$

(Notice that since f is continuous at b, we dropped the requirement that $0 < |z - b|$.) But since $z = g(x)$, $limit\ g(x) = b$, and δ is a positive real number, the generalized limit definition implies there is a δ-set Δ such that

$$|g(x) - b| < \delta \qquad \text{if} \qquad x \in \Delta.$$

Hence, reversing the steps and using $z = g(x)$, we have

if $\qquad\qquad\qquad\qquad\qquad x \in \Delta,$

then $\qquad\qquad\qquad\quad |z - b| = |g(x) - b| < \delta;$

but then $\qquad\qquad\quad |f(z) - L| = |f(g(x)) - L| < \epsilon.$

We have proved that for every $\epsilon > 0$ there is a δ-set Δ (determined as Δ above) such that

$$|f(g(x)) - L| < \epsilon \qquad \text{if} \qquad x \in \Delta;$$

that is, $limit\ f(g(x)) = L = f(b)$. Also, since $b = limit\ g(x)$, $limit\ f(g(x)) = f(limit\ g(x))$. Hence, we have established the next theorem:

Theorem 3-5c *Let f of g be the composite of two functions f and g. If $limit\ g(x)$ $= b$ and $\underset{b}{lim}\ f(z) = f(b) = L$ (i.e., f is continuous at b), then $limit\ f(g(x))$ $= L = f(b) = f(limit\ g(x))$. Traditionally, we write $\underset{x \to a}{lim}f(g(x)) = \underset{g(x) \to b}{lim}f(g(x))$.*

Example 3 Determine $\underset{\to}{lim}\cos(1/x)$ by Theorem 3–5c.

Solution: Since $\lim_{\to} (1/x) = 0$ and $\lim_{0} \cos z = \cos 0 = 1$, then $\lim_{\to} \cos (1/x) = 1$ by Theorem 3–5c.

As a special case of Theorem 3–5c, we have the following theorem concerning continuity of functions.

Theorem 3-5d *Let f of g be the composite of two functions f and g. If g is continuous at a and f is continuous at g(a), then f of g is continuous at a.*

Proof: Since g is continuous at a, $\lim_{a} g(x) = g(a)$. Since f is continuous at $g(a)$, $\lim_{g(a)} f(z) = f(g(a))$. Therefore, $f(g(a))$ is defined, and by Theorem 3–5c $\big($with $g(a) = b\big)$

$$\lim_{a} f(g(x)) = f(g(a));$$

that is, f *of* g is continuous at a.

Example 4 Show that the function whose equation is $y = \sqrt{x^2 + 1}$ is continuous at b for every real number b.

Solution: Let b be any real number. Since $\lim_{b} (x^2 + 1) = b^2 + 1$, $b^2 + 1 > 0$, and $\lim_{b^2+1} \sqrt{z} = \sqrt{b^2 + 1}$; then $\lim_{b} \sqrt{x^2 + 1} = \sqrt{b^2 + 1}$ by Theorem 3–5d. Hence, the function is continuous at b for every real number b.

Exercises

For Exercises 1 through 3, prove that the functions whose equations are the following are each continuous at the real number b.

1. $y = 3x^2 - 2x$. Let b be any real number.

2. $y = \dfrac{x^2 - 1}{x^2 + 1}$. Let b be any real number.

3. $y = (3/x^2) - (5/x) + 7$. Let b be any real number except 0.

For Exercises 4 through 6, we have proved that $\lim_{b} \sqrt{x} = \sqrt{b}, b > 0$ (Example 4 of §2–14). Hence, $f = \{(x, \sqrt{x})\}$ is continuous at b for every positive real number b. Use Theorem 3–5c to determine the following limits:

4. $\lim_{2} \sqrt{x - 1}$.

5. $\lim_{\to} \sqrt{(1 + x)/x}$.

6. $\lim_{0} \sqrt{(2/x) \sin x}$.

For Exercises 7 and 8, prove that each of the following functions is continuous at b where b is any real number.

7. $F = \{(x, \sin(x - b))\}.$ **8.** $G = \{(x, \cos(x - b))\}.$

9. Prove that $\sin = \{(x, \sin x)\}$ is continuous at b for every real number b by completing the following discussion:

$$\sin x = \sin((x - b) + b) = \sin b \cos(x - b) + \cos b \sin(x - b).$$

Hence, by Exercises 7 and 8, $\lim_{b} \sin x = \; ????$

10. (Optional) For an example to show the importance of the requirement in Theorem 3–5c that f be continuous, consider the following functions f and g. Note that f has a limit at 0 but f is *not* continuous at 0.

$$g(x) = \begin{cases} x & \text{if } x \geq 0, \\ 0 & \text{if } x < 0. \end{cases} \qquad f(z) = \begin{cases} 2 & \text{if } z = 0, \\ 1 & \text{if } z \neq 0. \end{cases}$$

Determine that **(a)** $\lim_{0} g(x) = 0$; **(b)** $\lim_{0} f(z) = 1 \neq f(0)$; **(c)** $\lim_{0} f(g(x))$ does not exist.

3-6 Limits in High School Mathematics

Several topics frequently found in high school mathematics curricula may be presented with discussions that are dependent upon limits. Generally, the discussions involve little more than a cursory description of a limit and contain arguments that rely only on intuition. Some of the more common topics are:

  (i) Circumference of a circle,
  (ii) Area of a circle,
  (iii) Unending decimal expressions,
  (iv) Completeness property,
  (v) Asymptotes of a hyperbola whose equation $\dfrac{x^2}{a^2} - \dfrac{y^2}{b^2} = 1$, $a \neq 0$ and $b \neq 0$,
  (vi) Sum of an infinite geometric series.

Of the topics above, we have discussed all except (ii), (v), and (vi) and we shall consider these topics here. Since (i) and (ii) are related and are dependent, we consider both together:

(i), (ii) *Circumference and area of a circle:* We assume, as was suggested in §1–5, that $\lim p_n = C$ where p is a sequence of perimeters of regular polygons of n sides inscribed in a given circle (let $p_1 = p_2 = 0$ when $n = 1$, $n = 2$). A special sequence of perimeters of inscribed regular polygons of $4, 8, 16, \ldots, 2^{n+1}, \ldots$ sides was generated and discussed in §1–5.

To define π let r, p_n and r', p'_n be radii and perimeters, respectively, of two n-sided regular polygons inscribed in any two circles O and O'. By similar triangles it can be proved that for each n,

$$\frac{p_n}{r} = \frac{p_n'}{r'}.$$

But since the limit of a sequence is unique

$$lim \frac{p_n}{r} = lim \frac{p_n'}{r'}.$$

Hence, if C and C' are the circumferences of circles O and O', respectively, then

$$\frac{C}{r} = \frac{1}{r} lim\, p_n = lim\, \frac{p_n}{r} = lim\, \frac{p_n'}{r'} = \frac{1}{r'} lim\, p_n' = \frac{C'}{r'}.$$

Thus, we have shown that the ratio (C/r) for any two circles is constant. We call this constant 2π; then

$$\frac{C}{2r} = \pi \qquad \text{and} \qquad C = 2\pi r.$$

Given any circle, we define the area A of a circle in an analogous fashion: $A = lim\, A_n$ where A_n is the area of an n-sided regular polygon inscribed in the circle. To verify that $A = \pi r^2$, we evaluate the three limits in Examples 1 through 3: $lim\, s_n$, $lim\, a_n$, and then $lim\, A_n$ where $s_n, a_n,$ and A_n are the length of a side, the length of the apothem, and the area, respectively, of an n-sided regular polygon inscribed in a circle. The following figure provides convenient reference to these symbols in addition to geometrical relationships between the entities they represent.

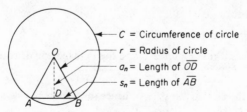

C = Circumference of circle
r = Radius of circle
a_n = Length of $\overline{OD}$
s_n = Length of $\overline{AB}$

Figure 3-5

Example 1 Prove that $lim\, s_n = 0$.

Solution: Let any $\epsilon > 0$ be given. To prove that $lim\, s_n = 0$ we must determine a value of δ, say δ', such that

$$-\epsilon < s_n < \epsilon \qquad \text{if} \qquad n > \delta'.$$

With $lim\, p_n = C$ there exists $\delta_1 > 0$ such that if $n > \delta_1$, then

$$C - \epsilon < p_n < C + \epsilon.$$

But since $ns_n = p_n$ and $n > 0$, this inequality is equivalent to the inequalities

$$C - \epsilon < ns_n < C + \epsilon,$$

and
$$\frac{C - \epsilon}{n} < s_n < \frac{C + \epsilon}{n}. \tag{1}$$

We can ensure that $s_n < \epsilon$ for every $n > \delta'$ by selecting $\delta' = \max \left[\delta_1, \frac{C + \epsilon}{\epsilon} \right]$. Thus, if $n > \delta'$, then $n > \delta_1$ and by (1) $s_n < \frac{C + \epsilon}{n}$; and if $n > \delta'$, then $n > \frac{C + \epsilon}{\epsilon}$ which yields $\frac{C + \epsilon}{n} < \epsilon$. Therefore, if $n > \delta'$,

$$s_n < \frac{C + \epsilon}{n} < \epsilon. \tag{2}$$

We know that since $n \geq 1$ and $C > 0$,

$$\epsilon \geq \frac{\epsilon}{n}, \quad -\epsilon \leq -\frac{\epsilon}{n}, \quad \text{and} \quad -\epsilon < \frac{C}{n} - \frac{\epsilon}{n} = \frac{C - \dot{\epsilon}}{n}$$

for every natural number n. It follows from (1) that if $n > \delta'$, then $n > \delta_1$

and
$$-\epsilon < \frac{C - \epsilon}{n} < s_n. \tag{3}$$

We have found $\delta' = \max \left(\delta_1, \frac{C + \epsilon}{\epsilon} \right)$ such that if $x > \delta'$, then inequalities (2) and (3) both hold and, consequently, $-\epsilon < s_n < \epsilon$. Thus, for every $\epsilon > 0$ there exists $\delta > 0$ (determined as δ') such that

$$0 - \epsilon < s_n < 0 + \epsilon \quad \text{if} \quad n > \delta;$$

that is, $\lim s_n = 0$.

Example 2 Prove that $\lim a_n = r$.

Solution: From Figure 3–5 and the Pythagorean Theorem, we obtain

$$a_n = \sqrt{r^2 - (\tfrac{1}{2}s_n)^2}.$$

We have shown that $\lim s_n = 0$; hence, by Theorem 3–4c and 3–4d

$$\lim (\tfrac{1}{2}s_n)^2 = \lim(\tfrac{1}{4}s_n^2) = \tfrac{1}{4} \cdot 0 \cdot 0 = 0.$$

Since r is a constant, $\lim r^2 = r^2$. Therefore, by Theorem 3–4b

$$\lim \left(r^2 - (\tfrac{1}{2}s_n)^2 \right) = r^2 - 0 = r^2.$$

We note that the "square root function" $f = \{(z, \sqrt{z}\,)\}$ is continuous at $z = r^2$ for every real number $r^2 > 0$. Therefore, by Theorem 3–5c

$$\lim a_n = \lim \sqrt{r^2 - (\tfrac{1}{2}s_n)^2} = \sqrt{r^2} = r.$$

Example 3 Prove that $\lim A_n = \pi r^2$.

Solution: From Figure 3–5 we see that triangle OAB has area $\frac{1}{2}s_na_n$. Hence,

$$A_n = \tfrac{1}{2}ns_na_n = \tfrac{1}{2}p_na_n.$$

Since $lim\, p_n = C = 2\pi r$ and $lim\, a_n = r$, then by Theorems 3–4c and 3–4d,

$$lim\, A_n = \tfrac{1}{2}(2\pi r)r = \pi r^2.$$

Thus, when we assume that $lim\, p_n = C$ and $lim\, A_n = A$, we can conclude that $A = \pi r^2$ which was to be proved.

(v) *Asymptotes of a hyperbola whose equation is* $\dfrac{x^2}{a^2} - \dfrac{y^2}{b^2} = 1, a \neq 0$ *and* $b \neq 0$: To use limits of functions to discuss the graph of this equation, let us solve the equation for y to obtain

$$y = \frac{b}{a}\sqrt{x^2 - a^2} \qquad \text{or} \qquad y = -\frac{b}{a}\sqrt{x^2 - a^2}.$$

The equation $y = (b/a)\sqrt{x^2 - a^2}$ can be used to define a function f whose graph is a part of the hyperbola in the first quadrant of the Cartesian coordinate system:

$$f = \left\{ \left(x, \frac{b}{a}\sqrt{x^2 - a^2} \right) \right\}, \qquad x \geq |a|.$$

A line whose equation is $y = l(x)$ is a **right-hand non-vertical asymptote** of a function f iff $\underset{\rightarrow}{lim}\,|f(x) - l(x)| = 0$. Similarly, the line is a **left-hand non-vertical asymptote** iff $\underset{\leftarrow}{lim}\,|f(x) - l(x)| = 0$. We shall prove that the line whose equation is $y = (b/a)x$ is a right-hand non-vertical asymptote of f by proving that

$$\underset{\rightarrow}{lim} \left(\frac{b}{a}\sqrt{x^2 - a^2} - \frac{b}{a}x \right) = 0.$$

First note that

$$\frac{b}{a}\sqrt{x^2 - a^2} - \frac{b}{a}x = \frac{b}{a}(\sqrt{x^2 - a^2} - x)$$

$$= \frac{b}{a}\frac{-a^2}{\sqrt{x^2 - a^2} + x}$$

$$= \frac{-ba}{\sqrt{x^2 - a^2} + x}.$$

Thus, our task will be completed when we prove that

$$\underset{\rightarrow}{lim} \frac{-ba}{\sqrt{x^2 - a^2} + x} = 0. \tag{4}$$

To prove this limit is 0, let any $\epsilon > 0$ be given. We consider only those real numbers x in the domain of the function; that is, $x \geq |a|$. Hence,

$$\left| \frac{1}{\sqrt{x^2 - a^2} + x} - 0 \right| = \frac{1}{\sqrt{x^2 - a^2} + x} \leq \frac{1}{x}$$

and $(1/x) < \epsilon$ provided $x > (1/\epsilon)$. Thus for any $\epsilon > 0$ we have found $\delta = (1/\epsilon)$ such that

$$\left| \frac{1}{\sqrt{x^2 - a^2} + x} - 0 \right| < \epsilon \qquad \text{if} \qquad x > \delta;$$

that is, $\lim\limits_{\to} \dfrac{1}{\sqrt{x^2 - a^2} + x} = 0$. Hence, equation (4) follows by Theorem 3–4c, and

$$\lim_{\to} \frac{-ba}{\sqrt{x^2 - a^2} + x} = -ba \cdot 0 = 0$$

which was to be shown.

In an analogous fashion the non-vertical asymptotes in the other quadrants can be determined and verified.

(vi) *Sum of an infinite geometric series:* An infinite geometric series is defined by the expression

$$a + ar + ar^2 + ar^3 + \cdots + ar^{n-1} + \cdots$$

where a and r are real numbers and $a \neq 0$ (r is frequently called the *common ratio*). To evaluate an infinite series consider a sequence of partial sums as defined in §1–12:

$$a, a + ar, a + ar + ar^2, \ldots, a + ar + ar^2 + \cdots + ar^{n-1}, \ldots$$

The sum S of the infinite geometric series is the limit of the sequence of partial sums. To determine the sum of the infinite series or the limit of the sequence of partial sums, we consider the cases where $r = 1$, $|r| < 1$, and $r = -1$ in Examples 4, 5, and 6, respectively, and the case where $|r| > 1$ in Exercises 2 and 3.

Example 4 Prove the series has no sum if $r = 1$.

Solution: If $r = 1$, the series is $a + a + \cdots + a + \cdots$ whose sequence of partial sums has the general term na. For $a > 0$, $na > M$ if $n > M/a$ where M is any positive real number; and for $a < 0$, $na < N$ if $n > N/a$ where N is any negative real number. Hence if $r = 1$ (and $a \neq 0$), the sequence of partial sums is unbounded (above when $a > 0$, below when $a < 0$), and therefore the sequence has no limit and the series has no sum.

When $r \neq 1$ we have for the general term of the sequence of partial sums

$$a + ar + ar^2 + \cdots + ar^{n-1} = \frac{a(1 - r^n)}{1 - r}$$

(see Exercise 1). Hence, from the theorems proved in §3–4 on limits of sums, differences, products, and quotients we can evaluate the limit of the sequence as follows: $lim\ a = a$, $lim\ 1 = 1$, $lim\ r = r$, and if $lim\ r^n = L$, then

$$lim\ \frac{a(1 - r^n)}{1 - r} = \frac{a(1 - L)}{1 - r}.$$

Thus, to find the sum of the infinite geometric series when $r \neq 1$, we need only to determine $lim\ r^n$ or prove that $lim\ r^n$ does not exist.

Example 5 (a) Prove that $lim\ r^n = 0$ if $|r| < 1$, and (b) then find S.

Solution: (a) If $r = 0$, then $lim\ r^n = 0$. If $0 < |r| < 1$, then $|r| = \dfrac{1}{1 + p}$ where p is a positive real number $\left(p = \dfrac{1 - |r|}{|r|}\right)$. Using this expression for $|r|$, the binomial theorem, and laws of exponents, we can write the equalities

$$|r^n| = |r|^n = \left(\frac{1}{1 + p}\right)^n = \frac{1}{(1 + p)^n}$$

$$= \frac{1}{1 + np + \dfrac{n(n - 1)}{2}p^2 + \cdots + p^n}.$$

But

$$\frac{1}{1 + np + \dfrac{n(n - 1)}{2}p^2 + \cdots + p^n} \leq \frac{1}{np}.$$

Thus, $|r^n| \leq \dfrac{1}{np}$, and it follows that $|r^n| \leq \dfrac{1}{np} < \epsilon$ if $n > \dfrac{1}{p\epsilon}$. Therefore, we have found a value of δ, namely $\delta = 1/p\epsilon$, as required by the limit definition, and we conclude that $lim\ r^n = 0$ if $0 < |r| < 1$. (b) Now the sum of the infinite geometric series can be determined. If $|r| < 1$, then $lim\ r^n = 0$ and

$$S = lim\ \frac{a(1 - r^n)}{1 - r} = \frac{a(1 - 0)}{1 - r} = \frac{a}{1 - r}.$$

Example 6 Prove that the infinite geometric series has no sum if $r = -1$.

Solution: If $r = -1$, then the series is $a - a + a - a + a - a \cdots$ and the sequence of partial sums is $a, 0, a, 0, \ldots, a, 0, \ldots$. But this sequence has no limit (see Example 2 of §1–10) and therefore the infinite series has no sum.

Upon completing Exercises 2 and 3 we shall have considered all values of r. Then we shall be able to conclude that an infinite geometric series has a sum S iff $|r| < 1$, and that sum is $\dfrac{a}{1 - r}$.

Summary If our proofs of these traditional high school topics seem quite involved and complicated, it is only because we have treated the topics in detail. Informal discussions such as, "the length of a side s_n approaches zero as n increases without bounds," may seem pedagogically more desirable provided no student thinks to ask, "Why?" Then the limit definitions and theorems that we have given are very appropriate.

Some exercises are provided to permit you to become acquainted with some other uses of the limit concept in high school mathematics. Other exercises are to be worked to complete the discussion of the sum of an infinite geometric series.

Exercises

1. Let $s_n = a + ar + ar^2 + \cdots + ar^{n-1}$. (a) Determine $s_n - rs_n$ (b) Express s_n in terms of a, r, and n.

2. Prove that if $|r| > 1$ then $\lim r^n$ does not exist. *Hint:* Assume that $\lim r^n = L$ where L is some real number and $|r| > 1$. Recall that we have proved that $\lim (1/r)^n = 0$ when $|(1/r)| < 1$. Then consider $\lim [(r)^n(1/r)^n]$.

3. Let $|r| > 1$ and $s_n = \dfrac{a(1 - r^n)}{1 - r}$. Since $\lim r^n$ does not exist (see Exercise 2), prove $\lim s_n$ does not exist. *Hint:* If $s_n = \dfrac{a(1 - r^n)}{1 - r}$, then $r^n = 1 - \dfrac{(1 - r)s_n}{a}$. Assume $\lim s_n = L$ and seek a contradiction.

4. Let a be an infinite series and s be a sequence of partial sums such that for each natural number n

$$s_{n-1} = a_1 + a_2 + a_3 + \cdots + a_{n-1}$$

and

$$s_n = a_1 + a_2 + \cdots + a_{n-1} + a_n.$$

Then $s_n - s_{n-1} = a_n$. Prove that $\lim a_n = 0$ if $\lim s_n = L$; that is, prove that the sequence of terms of the series has limit 0 if the series has a sum.

5. Given the area of a circle $A = \pi r^2$, a limit can be used to determine the circumference of the circle as follows: Consider two concentric circles with radii r and $r + h$ ($h > 0$), respectively, in a plane. The area of the "band" between the circles is $\pi(r + h)^2 - \pi r^2$. But this band, if "unrolled" would be, approximately, a rectangle whose area is found by multiplying length of the base b times the height h. Hence, b is a function of h, $b = \dfrac{\pi(r + h)^2 - \pi r^2}{h}$, and we can define C, the circumference of the circle whose radius is r, to be $\lim_{0^+} \dfrac{\pi(r + h)^2 - \pi r^2}{h}$. Find C.

6. A discussion similar to that in Exercise 5 can be given to determine the surface area of a sphere if the volume is known. Let the volume of the

"shell" between two concentric spheres be $\frac{4}{3}\pi(r + h)^3 - \frac{4}{3}\pi r^3$ where $h > 0$. Let the surface area S of a sphere with radius r be defined to be $\lim\limits_{0^+} \frac{\frac{4}{3}\pi(r + h)^3 - \frac{4}{3}\pi r^3}{h}$, and find S.

3-7 On from Here . . .

The generalized limit definition and the generalized limit theorems apply directly to types of limits other than the six we have discussed. One example should help us to realize this fact and to recognize some other examples.

Example 1 Let $f(x, y) = z$ be the equation of a real function of two variables whose domain is the set of all ordered pairs of real numbers (x, y). The function f will be a set $\{((x, y), z)\}$ such that for each pair of real numbers (x, y) there exists one and only one real number z. We say that f has a **limit at** (a, b) equal to L, $\lim\limits_{(a, b)} f(x, y) = L$, iff for every real number $\epsilon > 0$ there exists a real number $\delta > 0$ such that

$$|f(x, y) - L| < \epsilon \quad \text{if} \quad 0 < \sqrt{(x - a)^2 + (y - b)^2} < \delta.$$

Show that the generalized limit definition applies to this specific limit.

> *Solution:* The notation "$|f(x) - L| < \epsilon$" in the generalized limit definition must be extended to include $|f(x, y) - L| < \epsilon$. However, recall that "x" was a symbol for an element in the domain of the function under question, and it should seem natural for "x", in the generalized limit definition, to be the symbol for any domain element. Therefore, the expression $|f(x, y) - L| < \epsilon$ conforms to the generalized limit definition.
>
> The question of primary concern is whether the truth set S of the inequality $0 < \sqrt{(x - a)^2 + (y - b)^2} < \delta$ (the set of ordered pairs of real numbers that satisfy the inequality) comprises what we would call a δ-set. Note that the graph of S is the set of points inside the circle whose equation is
>
> $$(x - a)^2 + (y - b)^2 = \delta^2$$
>
> except for the center. Let us call such a set as S a **deleted circular neighborhood of** (a, b), and consider the properties of δ-sets.
>
> **Property 1 of δ-sets** *If the domain D_f of f is the set of all ordered pairs of real numbers or any subset that contains a single deleted circular neighborhood of (a, b), then D_f contains at least one deleted circular neighborhood of (a, b).*
>
> **Property 2 of δ-sets** *If $\delta > 0$, then deleted circular neighborhoods of (a, b) are non-empty sets.*

Property 3 of δ-*sets* If δ_1 and δ_2 are two positive real numbers, the intersection of the two deleted circular neighborhoods of (a, b) is that one deleted circular neighborhood that has the least radius. If (x, y) is in a circle whose center is (a, b) and whose radius is δ where $\delta = min\,[\delta_1, \delta_2]$, then (x, y) is in both circles whose centers are (a, b) and whose radii are δ_1 and δ_2, respectively.

Therefore, we have shown that the domain variable can be interpreted as a symbol for ordered pairs of real numbers, and we have proved that the truth sets of the inequalities $0 < \sqrt{(x-a)^2 + (y-b)^2} < \delta$ are δ-sets. (We call them δ-sets because they satisfy properties for δ-sets.) Hence, the generalized limit definition applies to this particular limit definition as it does to the six limit definitions discussed in §3–3.

The solution of Example 1 was somewhat lengthy. However, was it worth our time and effort? We did define a different type of limit: a limit of a function of two variables. Also, we verified that the generalized limit definition applies to this new particular limit and that the truth sets of the condition $0 < \sqrt{(x-a)^2 + (y-b)^2} < \delta$ satisfy the properties for δ-sets. If $\lim_{(a,\,b)} f(x, y) = L$, can there be a $\lim_{(a,\,b)} f(x, y) = M$ where $L \neq M$? If g is another real function of two variables and $\lim_{(a,\,b)} g(x, y) = M$, is $\lim_{(a,\,b)} (f(x, \ y) + g(x, y)) = L + M$? An attempt to answer these questions by proving some theorems should be very enlightening—we would find that we have already proved generalized limit theorems that apply directly to this new type of limit. A proof of the theorem, "If $\lim_{(a,\,b)} f(x, y) = L$ and $\lim_{(a,\,b)} g(x, y) = M$, then $\lim_{(a,\,b)} (f(x, y) + g(x, y)) = L + M$," is a special case of the proof of Theorem 3–4a; the proof can be formed by changing "$f(x)$" and "$g(x)$" to "$f(x, y)$" and "$g(x, y)$", respectively, and by changing references to δ-sets to involve the inequality $0 < \sqrt{(x-a)^2 + (y-b)^2} < \delta$. Thus, by showing that the generalized limit definition and properties of δ-sets pertain to this new particular limit, we find that all of the generalized limit theorems apply to this new particular limit and we need not reprove the theorems.

In general, if we ever find a limit definition to which the generalized limit definition and properties of δ-sets apply, then we are assured that all of the theorems proved in this chapter apply to the limit. This is true because the proofs of the generalized limit theorems depend upon the generalized limit definition and the properties of δ-sets. (We make the sweeping generalization under the assumption that δ, L, and $f(x)$ are real numbers and that the algebraic and order properties of the real numbers are to be used.)

The concepts involved in the formation of the generalized limit definition and theorems apply to other systems (or spaces) also. One of many possible extensions of the generalized limit concept is found in the next example.

Example 2 Consider the definition: A real function $f = \{(x, f(x))\}$ **increases without bound at-the-right** iff for every real number M there exists some real number $\delta > 0$ such that

$$f(x) > M \quad \text{if} \quad x > \delta.$$

A definition of a function f that increases without bound at-the-left, or at a real number b, and so forth can be made by changing "$x > \delta$" to "$x < -\delta$" or to "$0 < |x - b| < \delta$," and so forth. Form a generalized definition of a function f that increases without bound on some δ-set.

> *Solution:* A function $f = \{(x, f(x))\}$ increases without bound on some δ-set iff for every real number M there exists some $x \in \Delta$ such that
>
> $$f(x) > M \quad \text{if} \quad x \in \Delta.$$
>
> If the δ-sets of §3–3 are considered, this definition gives each of the definitions suggested in the statement of Example 2.

Finally, let us mention that the generalized limit concept introduced in Example 2 can be generalized to an even greater extent. A function f that increases without bound is frequently said to have *limit infinity* and is expressed as $\lim f(x) = \infty$ or $\lim_{x \to \infty} f(x) = \infty$. A more general generalized limit definition can be formed to include even such "limits" as "infinity." However, we shall not undertake such a task. If you have proceeded successfully this far in this book you should be quite capable of reading more advanced texts for a more general, more advanced point of view. (See modern texts in calculus, topology, and functions of a real or complex variable.) Also, if you have been a participating reader, you may be able to form extensions of the limit theory that we have discussed to include limit infinity, limits of geometrical configurations, limiting positions of secant lines of a circle drawn through a common point, and many other notions that might be original to you. The following exercises may disclose some possible extensions of the limit concept that we have studied.

Exercises

1. Prove that $f = \{(x, 2x)\}$ increases without bound at-the-right.

2. Prove that $g = \{(x, (1/x^2))\}$, $x \neq 0$ increases without bound at 0; that is, prove that for every real number M there exists a real number $\delta > 0$ such that $(1/x^2) > M$ if $0 < |x - 0| < \delta$.

3. Let A and B be ends of a diameter of any circle O. Consider a set of points p_n on the circle that satisfy the following properties: p_1 is B; p_2 bisects the arc $\overparen{p_1 A}$; p_3 bisects the arc $\overparen{p_2 A}$; in general p_n bisects the arc $\overparen{p_{n-1} A}$. What is the "limiting position" of the ordered set of points p_1, $p_2, p_3, \ldots, p_n, \ldots$?

4. Consider the ordered set of points $p_1, p_2, \ldots, p_n, \ldots$ in Exercise 3 and imagine secant lines drawn through A and p_n for every n; that is, consider the ordered set of secants $Ap_1, Ap_2, \ldots, Ap_n, \ldots$. What is the "limiting position" of this ordered set of secant lines?

3-8 Overview

We have studied in detail six types of limits; in abbreviated notation they are $\lim_{\rightarrow}, \lim_{\leftarrow}, \lim_{b^+}, \lim_{b^-}, \lim_{b}$. In each case, various examples were evaluated, and a few theorems were proved through the use of each limit definition. We have become aware of a close resemblance among the six limit definitions and therefore have been able to extend our knowledge of a particular limit to studies of other types of limits.

The single most significant development in our study has been that of the generalized limit. By comparing and interrelating the six limit definitions we have been able to determine a definition of a generalized limit and to characterize δ-sets that enable us to apply the generalized limit concept to each of the six previously discussed types of limits. Because a proof of a generalized limit theorem depends only upon the generalized limit definition and properties of the δ-sets, each proof of a generalized limit theorem effects a proof of six individual theorems for six types of limits. What is even more important, such a proof is valid for other limits, such as $\lim_{(a, b)} f(x, y) = L$, provided it can be interpreted to satisfy the generalized limit definition and properties of δ-sets. Therefore, through the generalized limit theorems we have proved a large quantity (over one hundred) particular theorems that apply not only to limits that have been discussed but to limits yet to be encountered.

In addition to applying the limit concept that we have discussed we could have extended or revised the concept. We required that the domain D_f of a function f contain certain real numbers; that is, D_f was to contain a neighborhood of b for a limit at b, an open right ray for a limit at-the-right, and so forth (see Figure 3–7 of §3–2). Also, when we proved by definition of a limit that a certain limit existed, we assumed that the statement

$$\text{"}|f(x) - L| < \epsilon \quad \text{if} \quad x \in \Delta\text{"}$$

inferred that if $x \in \Delta$, then $x \in D_f$. Following are two definitions that a student may find in the literature. These two apply to limits of a real sequence or function and differ from our definitions primarily in the conditions on the domain D_f of a function f. We state each definition in the language of the generalized limit:

(i) Let f be a (real) function whose domain D_f contains an element in every δ-set. Then $\lim it f(x) = L$ iff for every $\epsilon > 0$ there exists a δ-set Δ such that

$$|f(x) - L| < \epsilon \quad \text{if} \quad x \in (\Delta \cap D_f).$$

(ii) Let f be a (real) function. Then *limit* $f(x) = L$ iff for every $\epsilon > 0$ there exists a δ-set Δ such that

$$|f(x) - L| < \epsilon \quad \text{if} \quad x \in \Delta.$$

Although we have not studied the results of these definitions, we should be able to do so readily.

As discussed in §3-6, concepts frequently found in the high school mathematics curricula involve limits. In addition to these topics, high school and college mathematics students and their teachers are, to a varying extent, involved with limits in calculus.

The central concept in calculus is the concept of limits that we have discussed. For instance, let f be a function whose domain contains a neighborhood of b. Then f has a **derivative at** b iff $\lim_{b} \frac{f(b) - f(x)}{b - x} = L$ where L is a real number. We write $L = f'(b)$ and call $f'(b)$ the **derivative of** f **at** b. Thus, given a function f and a real number b we may determine the **difference quotient function** F, $F(x) = \frac{f(b) - f(x)}{b - x}$ as shown in Figure 3-6, and then proceed to apply our definition, theorems, and understanding of the limit at b to $\lim_{b} F(x)$. A study of this limit, $\lim_{b} \frac{f(b) - f(x)}{b - x}$, is a study in **differential calculus.**

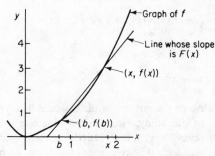

Let $f = \{(x, x^2)\}$

Then $F(x) = \frac{f(b) - f(x)}{b - x}$

Hence, $f'(b) = \lim_{b} F(x)$

$= \lim_{b} \frac{b^2 - x^2}{b - x}$

$= \lim_{b} (b + x)$

$= 2b$

Figure 3-6

Let f be a bounded function whose domain contains a closed interval $[a, b]$, and let $x_0, x_1, x_2, \ldots, x_n$ each be real numbers in $[a, b]$ such that for every $i(i = 0, 1, 2, \ldots, n)$

$$a = x_0 \leq x_{i-1} < x_i \leq x_n = b,$$

$$x_i - x_{i-1} = \frac{b - a}{n},$$

m_i is the greatest lower bound of f on $[x_{i-1}, x_i]$,

and M_i is the least upper bound of f on $[x_{i-1}, x_i]$.

Then for each value of n, unique real numbers s_n and S_n,

$$s_n = \left(\frac{b-a}{n}\right)[m_1 + m_2 + m_3 + \cdots + m_n],$$

$$S_n = \left(\frac{b-a}{n}\right)[M_1 + M_2 + M_3 + \cdots + M_n],$$

can be computed to generate two sequences $\{(n, s_n)\}$ and $\{(n, S_n)\}$ called a **lower Riemann sum** and an **upper Riemann sum**, respectively. Then if $lim\, s_n = lim\, S_n = L$ where L is a real number, we write $L = \int_a^b f(x)\, dx$, and we call $\int_a^b f(x)\, dx$ the **definite integral** of f on $[a, b]$. Thus, given a function f and an interval $[a, b]$ we may compute a sequence s of lower Riemann sums, and a sequence S of upper Riemann sums, as shown in Figure 3-7 and apply our definition, theorems, and understanding of the limit of a sequence to $lim\, s_n$ and $lim\, S_n$. A study of these limits is a study in **integral calculus**.

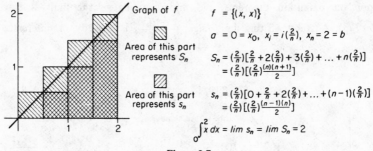

$f = \{(x, x)\}$

$a = 0 = x_0, \; x_i = i\left(\frac{2}{n}\right), \; x_n = 2 = b$

Area of this part represents S_n

$S_n = \left(\frac{2}{n}\right)[\frac{2}{n} + 2(\frac{2}{n}) + 3(\frac{2}{n}) + \ldots + n(\frac{2}{n})]$
$= \left(\frac{2}{n}\right)[(\frac{2}{n})\frac{(n)(n+1)}{2}]$

Area of this part represents s_n

$s_n = \left(\frac{2}{n}\right)[0 + \frac{2}{n} + 2(\frac{2}{n}) + \ldots + (n-1)(\frac{2}{n})]$
$= \left(\frac{2}{n}\right)[(\frac{2}{n})\frac{(n-1)(n)}{2}]$

$\int_0^2 x\, dx = lim\, s_n = lim\, S_n = 2$

Figure 3-7

A rigorous study of the derivative or the integral, while it would involve limits, has not been the purpose of our study. Nor was our objective the evaluation of special sequences or series for π, e, ln 2, Euler's constant, and other known numbers, or the study of special convergence conditions, such as the Cauchy convergence condition. Rather, our objective has been to provide experiences enabling you, a participating reader, to grasp the basic conceptual notion of a limit. Now you should be prepared to make a study of calculus and other topics in analysis, understand or find for yourself derivations of special limits and convergence conditions, and appreciate the significant role that a concept of limits plays in mathematics.

Glossary of Symbols

Symbol	Meaning or Way to Read
$=$	Read, "is equal to," "which is equal to," or "be equal to."
$\neq$	Not equal to.
$<$	Read, "is less than," "which is less than," or "be less than."
$\leq$	Less than or equal to.
$>$	Read, "is greater than," "which is greater than," or "be greater than."
$\geq$	Greater than or equal to.
(a, b)	An ordered pair of numbers, coordinates of a point. Also, an open interval.
a, t, s, etc.	Infinite sequences, called "sequences."
$s_1, s_2, s_3, \ldots, s_n, \ldots$	A sequence called "s."
$\{(n, s_n)\}$	A sequence called "s."
$\{s_1, s_2, s_3, \ldots, s_n, \ldots\}$	The range of a sequence s.
iff	Read, "if and only if."
$\overline{AB}$	Segment AB or the length of segment AB.
ϵ	Epsilon: a variable for a positive real number.
$|x|$	Absolute value of x.
δ	Delta: a variable for a positive real number.
$lim\, s_n, lim\{(n, s_n)\}$	The limit of a sequence s.

SYMBOL	MEANING OR WAY TO READ
f, g, h, F, G, H	Functions.
$f(x)$	The number associated with x by the function f. Read, "f of x."
D_f	The domain of the function f.
$\{(x, y)\}$	A function.
$\{(x, y) \mid y = f(x)\}$	Read, "the set of ordered pairs of real numbers x, y such that $y = f(x)$.
$\in$	Read "is an element of" or "which is an element of."
(a, b)	An open interval, $\{x \mid a < x < b\}$.
$[a, b]$	A closed interval, $\{x \mid a \leq x \leq b\}$.
$(a, \rightarrow)$	An open right ray, $\{x \mid a < x\}$.
$[a, \rightarrow)$	A closed right ray, $\{x \mid a \leq x\}$.
$(\leftarrow, b)$	An open left ray, $\{x \mid x < b\}$.
$(\leftarrow, b]$	A closed left ray, $\{x \mid x \leq b\}$.
$\lim_{\rightarrow} f(x), \lim_{\rightarrow} \{(x, f(x))\}$	The limit at-the-right of the function f.
$\lim_{\leftarrow} f(x), \lim_{\leftarrow} \{(x, f(x))\}$	The limit at-the-left of the function f.
$\lim_{b^+} f(x), \lim_{b^+} \{(x, f(x))\}$	The right-side limit at b of the function f.
$\lim_{b^-} f(x), \lim_{b^-} \{(x, f(x))\}$	The left-side limit at b of the function f.
$\lim_{b} f(x), \lim_{b} \{(x, f(x))\}$	The limit at b of the function f.
$\min [a, b]$	The minimum of the pair a and b.
$\max [a, b]$	The maximum of the pair a and b.
$A \cup B$	A union B.
$A \cap B$	A intersection B.
$f + g$	The sum of f and g.
$f - g$	The difference of f minus g.
$f \times g$	The product of f times g.
$f \div g$	The quotient of f divided by g.
f of g	The composite function f of g.
$\lim$	A type of limit, the limit of a sequence.
$\lim_{\rightarrow}$	A type of limit, the limit at-the-right.
$\lim_{\leftarrow}$	A type of limit, the limit at-the-left.

SYMBOL	MEANING OR WAY TO READ
$\lim\limits_{b^+}$	A type of limit, the right-side limit at b.
$\lim\limits_{b^-}$	A type of limit, the left-side limit at b.
$\lim\limits_{b}$	A type of limit, the limit at b.
δ-sets	Delta sets.
Δ	Capital delta, called "del."
$\subset$	Read, "is a subset of" or "which is a subset of."
$\overarc{AB}$	The arc AB or the measure of arc AB.

Bibliography

Buck, R. Creighton. *Advanced Calculus.* New York: McGraw-Hill Book Company, 1956.

Hight, Donald W. *The Limit Concept in the SMSG Revised Sample Textbooks.* Stillwater: Oklahoma State University Bookstore, 1961. (Multilithed).

McShane, E. J. "A Theory of Limits," *Studies in Modern Analysis.* (M.A.A. Studies in Mathematics, Vol. 1, R. C. Buck, editor.) Englewood Cliffs, N.J.: Prentice-Hall, Inc., 1962, pp. 7–29.

Moore, E. H. and Smith, H. L. "A General Theory of Limits," *Am. J. Math.*, **44** (1922), pp. 102–121.

Olmsted, John M. H. *The Real Number System.* New York: Appleton-Century-Crofts, 1962.

Randolph, John F. "Limits," *Insights Into Modern Mathematics*, Twenty-third Yearbook of the National Council of Teachers of Mathematics. Washington, D.C., 1957, pp. 200–240.

Smith, H. L. "A General Theory of Limits," *Natl. Math. Mag.* **12** (1938), pp. 371–379.

Answers to
Odd-Numbered Exercises

CHAPTER 1—SEQUENCES AND THEIR LIMITS

1-1 Infinite Sequences

1. 1, 3, 5, 7, 9.   **3. (a)** $\{2\}$. **(b)** $\{(n, 2)\}$.

5. Among the many correct answers are: 3, 2, 1, 1, 1, . . . where $s_n = 1$ for every $n \geq 3$; and 1, 2, 3, . . . , 1, 2, 3, . . . where $s_n = 1$ if $n = 3k - 2$, $s_n = 2$ if $n = 3k - 1$, $s_n = 3$ if $n = 3k$ where k is a natural number.

7. a, ar, ar^2, ar^3, ar^4.

1-2 Graphs of Infinite Sequences

1.

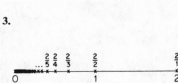

5.

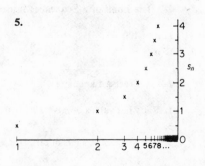

3.

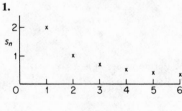

119

1-3 Some Examples to Ponder

1. (a) through (d) IIa, IIc, IId, IIIc, IIId. **3.** All of Group I, IIIa, IIIb.
5. All of Group I, IIIa, IIIb.   **7.** Ib, IIa, IIc, IId.

1-4 Types of Sequences

1. Ib, IIa, IIc, and IId are monotone increasing. Id and IIb are monotone decreasing. The others are not monotone.

3. (a) through (c) Yes. (d) Yes; every M where M $\geq \frac{1}{2}$. (e) $\frac{1}{2}$.

5. (a) through (c) Yes. (d) No; 1 is the greatest lower bound.

7. (a) A number m is the **greatest lower bound** of a sequence s iff m is a lower bound and $m \geq m'$ where m' is any lower bound of s. (b) A sequence s is **monotone decreasing** iff $s_{n+1} < s_n$ for every n. (c) A sequence s is **monotone nonincreasing** iff $s_{n+1} \leq s_n$ for every n.

1-5 Circumference of a Circle

1. (a) Bounded (above by $\frac{1}{2}\sqrt{2}$, below by 0). (b) Monotone decreasing.

3. No; $p_n < C$ for every n and $\pi = C$ when $d = 1$.

5. (a) Bounded (above by 4, the perimeter of a circumscribed square, and below by π, the circumference of the circle). (b) Monotone decreasing.

7. (a) Bounded (above by $\frac{1}{4}\pi$, the area of the circle, and below by $\frac{1}{2}$, the area of the inscribed square). (b) Monotone increasing.

1-6 Sequences in the Physical World and in History

1. All of the sequences are bounded.

3. The least upper bounds are as follows: for t, 1; for a and r, 2; for h, 1; for b, 3; and for d, 1.

5. (a) 0. (b) On the flat surface, at height 0.

1-7 The Limit of a Sequence: Informal

1. $n > 100$.   **3.** For no values of n.
5. $n > k$.   **7.** For no values of n.

1-8 A Precise Language

1. For what values of n is $|(1/n) - 0| < \epsilon$?

3. $n > 2/\epsilon$.

5. No; in Exercise 3 it is not true that $|(2/n) - 0| < \epsilon$ if $n > 1/\epsilon$.

1-9 Limit of a Sequence

1. Let $\delta = 1/\epsilon$; then for every $\epsilon > 0$ there exists $\delta > 0$ ($\delta = 1/\epsilon$) such that $|(1/n) - 0| < \epsilon$ if $n > \delta$; therefore, $\lim (1/n) = 0$.

3. Let $\delta = 2/\epsilon$; then for every $\epsilon > 0$ there exists $\delta > 0$ ($\delta = 2/\epsilon$) such that $|(2/n) - 0| < \epsilon$ if $n > \delta$; therefore, $\lim (2/n) = 0$.

5. (a) Yes. (b) Yes. (c) No. (d) Yes; if $\delta > 1/\epsilon$ and $n > \delta$, then $n > 1/\epsilon$ and $|(1/n) - 0| < \epsilon$.

1-10 Graphical Interpretation of the Limit

1. This set of points is a finite set; it includes only the points (n, s_n) when $n = 4, 5, 6, 7, 8,$ and 9.

3. (b) Cluster points are 1 and -1. (c) Ia, 0; Ib, 1; Ic, 0; Id, 1.

5. One and only one.

1-11 Theorems on Limits

1. Since $\lim s_n = L > 0$, then when $\epsilon = \frac{1}{2}L$ there exists some $\delta > 0$ (by the limit definition) such that $|s_n - L| < \frac{1}{2}L$ if $n > \delta$. But if $|s_n - L| < \frac{1}{2}L$, then $s_n > \frac{1}{2}L > 0$. Hence, there exists δ such that $s_n > 0$ for $n > \delta$.

3. We shall prove that $\lim u_n = L$ by determining $\delta_u > 0$ such that $|u_n - L| < \epsilon$ if $n > \delta_u$ when any $\epsilon > 0$ is given. Let $\epsilon > 0$ be given. Since $\lim s_n = L$, there exists $\delta_s > 0$ such that $|s_m - L| < \epsilon$ if $m > \delta_s$. Let $m = n + 4$, and since $u_n = s_{n+4} = s_m$ it follows that $|u_n - L| < \epsilon$ if $n + 4 > \delta_s$; that is, if $n > \delta_s - 4$. Thus choose $\delta_u = \delta_s$ then $\delta_u > 0$ and $n > \delta_s - 4$ if $n > \delta_u = \delta_s$. It follows that $|u_n - L| < \epsilon$ if $n > \delta_u$. Hence, $\lim u_n = L$.

5. We shall prove that $\lim t_n = L$ by determining $\delta_t > 0$ such that $|t_n - L| < \epsilon$ if $n > \delta_t$ when any $\epsilon > 0$ is given. Let $\epsilon > 0$ be given and consider $n > 3$. Since $\lim s_n = L$, there exists $\delta_s > 0$ such that $|s_m - L| < \epsilon$ if $m > \delta_s$. Now for $n > 3$ let $m = n - 3$; and since $t_n = s_{n-3} = s_m$, we are assured that $|t_n - L| < \epsilon$ if $n - 3 > \delta_s$; that is, if $n > \delta_s + 3$. Thus, choose $\delta_t = \delta_s + 3$; then it follows that $|t_n - L| < \epsilon$ if $n > \delta_t$. Hence, $\lim t_n = L$.

7. Let any $\epsilon > 0$ be given. Since $\lim s_n = L$, there exists $\delta > 0$ such that $L - \epsilon < s_n < L + \epsilon$ if $n > \delta$. But $L - \epsilon < b_n$ since $L < b_n$; and if $n > \delta$, then $b_n < L + \epsilon$ since $b_n < s_n$ and $s_n < L + \epsilon$. Therefore, $L - \epsilon < b_n < L + \epsilon$ if $n > \delta$. It follows that $\lim b_n = L$.

1-12 Applications of Limits of Sequences

1. (a) 1.0, 1.00, 1.000, $\ldots$, s_n, $\ldots$ where $s_n = 1$ (with n zeros if you wish). (b) Since this sequence is a constant sequence in which $s_n = 1$ for every n, $\lim s_n = 1$ (see Exercise 2 of 1–11).

3. (a) $0.9, 0.99, 0.999, \ldots, 1 - (1/10)^n, \ldots$ (b) Since $|(1 - (1/10)^n) - 1| = (1/10)^n$
and $(1/10)^n < 1/n$ for every natural number n, it follows that $|(1 - (1/10)^n)$
$- 1| < \epsilon$ if $n > 1/\epsilon$. Therefore, for every $\epsilon > 0$ there exists $\delta > 0$ $(\delta = 1/\epsilon)$
such that $|(1 - (1/10)^n) - 1| < \epsilon$ if $n > \delta$. Thus, $\lim (1 - (1/10)^n) = 1$.

5. $r_n < \sqrt{2}$ for every natural number n.

7. r_n is within $1/10^n$ of $\sqrt{2}$; that is, $|r_n - \sqrt{2}| < 1/10^n$.

CHAPTER 2—FUNCTIONS AND THEIR LIMITS

2-1 Functions

1. (a) $F(-3) = (-3)^2 = 9$. (b) $F(3) = 3^2 = 9$. (c) $F(5 - 7) = (5 - 7)^2 = 4$. (d)
$F(a + b) = (a + b)^2 = a^2 + 2ab + b^2$. (e) $F(a) + F(b) = a^2 + b^2$. (f) $F(a \cdot b)$
$= (ab)^2 = a^2 b^2$. (g) $F(a) \cdot F(b) = a^2 b^2$.

3. (a) It is true for some $(G(x) = 3x$ in Exercise 2), but not others $(F(x) = x^2$
in Exercise 1). (b) It is true for some $(F(x) = x^2)$, but not others $(G(x) = 3x)$.

5. $\{(x, y) \mid y = x\}$.

2-2 Graphs of Functions

1. The points in common are those whose coordinates can be expressed as
$\left(\frac{1}{2}\pi + 2n\pi, \dfrac{1}{\frac{1}{2}\pi + 2n\pi}\right), n = 0, 1, 2, \ldots$.

3. No. For every positive real number d we choose a natural number $n > d$;
then $(\frac{1}{2}\pi + n\pi) > d$, $\sin(\frac{1}{2}\pi + n\pi) = \pm 1$, and $\dfrac{\sin(\frac{1}{2}\pi + n\pi)}{\frac{1}{2}\pi + n\pi} = \dfrac{\pm 1}{\frac{1}{2}\pi + n\pi} \neq 0$.

5.

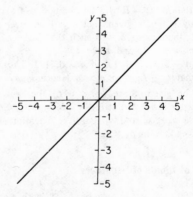

7. (a) $y = \pm 1$. (b) $y = \pm\frac{1}{2}\sqrt{2}$. (c) Associated with each value of x between -1
and 1, there are two (not one and only one) values of y.

2-3 Some Examples to Ponder

1. All of Groups I and III, and IIa.   **3.** Ia, Ic, IIa.
5. All of Group I, IIb, IIc, IId, IIIb, IIId.   **7.** All of Groups I and III.
9. Ib, IIb, IIc, IId.

2-4 Classification of Functions

1. (a) A function f is **bounded below** on a subset S of the domain of f iff there exists a real number M such that $f(x) \geq M$ for every x in S. (b) This number M is called a **lower bound** of f on S. (c) If a function is bounded above and bounded below on a subset S of the domain of f, f is said to be **bounded** on S. (d) A real number m is a **least upper bound** of a function f on a subset S of the domain of f iff m is an upper bound of f on S and $m \leq m'$ where m' is any upper bound of f on S. (e) A real number m is a **greatest lower bound** of f on a subset S of the domain of f iff m is a lower bound of f on S and $m \geq m'$ where m' is any lower bound of f on S. (f) A function f is **monotone decreasing** on a subset S of the domain of f iff for every pair of real numbers x_1, x_2 in S, $x_1 < x_2$ implies $f(x_1) > f(x_2)$. (g) Under the conditions of (f) on S, x_1, and x_2, if $f(x_1) \geq f(x_2)$ whenever $x_1 < x_2$, then we call f **monotone nonincreasing** on S. (h) A function f is **monotone** on a subset S of the domain of f iff f is monotone increasing, monotone decreasing, monotone nonincreasing, or monotone nondecreasing on S.

3. (a) Ib, Id. (b) Yes; the least upper bound is 1 for both functions.

5. (a) Ia, Id. (b) Both functions have a "limit" (as we shall determine in subsequent sections).

2-5 A Limit of a Function: Informal

1. (a) $x > 2$. (b) $x > 10$. (c) 0. (d) Yes. (A conclusion based on intuition. A definition will be given in the next section.)

3. No. Note that $|\sin x - 0| = 1$ if $x = \frac{1}{2}\pi + n\pi$ where $n = 0, 1, 2, 3, \ldots$.

5. $x > 1/\epsilon$.

2-6 The Limit At-The-Right

1. Let any $\epsilon > 0$ be given. Note that $|(1/x) - 0| = 1/x$ if $x > 0$, and $1/x < \epsilon$ if $x > 1/\epsilon$. Hence, let $\delta = 1/\epsilon$; then for every $\epsilon > 0$ there exists $\delta > 0$ ($\delta = 1/\epsilon$) such that $|(1/x) - 0| < \epsilon$ if $x > \delta$. The domain of the function contains $(0, \longrightarrow)$. Therefore, $\lim_{\longrightarrow} \{(x, 1/x)\} = 0$.

3. Let any $\epsilon > 0$ be given. Note that $\left| \dfrac{|x|}{x} - 1 \right| = 0$ if $x > 0$. Hence, let $\delta = 1$ (or any other positive real number); then for every $\epsilon > 0$ there exists $\delta > 0$ ($\delta =$

1) such that $\left|\dfrac{|x|}{x} - 1\right| < \epsilon$ if $x > \delta$. The domain of the function contains $(0, \rightarrow)$. Therefore, $\underset{\rightarrow}{lim}\ \dfrac{|x|}{x} = 1$.

5. (a) $|x - L| > 1$. (b) No. There is a real number x such that $x > \delta$ and $x > L + 1$ (e.g. $x = \delta + L + 1$). For this real number x, $|x - L| > 1$.

7. (a) $0 < y < 1$.   (b) $\underset{\rightarrow}{lim}\ \dfrac{1}{x} = 0,\ \underset{\rightarrow}{lim}\ \dfrac{x-1}{x} = 1$.

2-7 The Limit At-The-Left

1. Let any $\epsilon > 0$ be given. Note that $|(1/x) - 0| = -(1/x)$ if $x < 0$, and $-(1/x) < \epsilon$ if $x < -(1/\epsilon)$. Hence, let $\delta = 1/\epsilon$; then for every $\epsilon > 0$ there exists $\delta > 0$ $(\delta = 1/\epsilon)$ such that $|(1/x) - 0| < \epsilon$ if $x < -\delta$. The domain of the function contains $(\leftarrow, 0)$. Therefore, $\underset{\leftarrow}{lim}\ \{(x, 1/x)\} = 0$.

3. Let any $\epsilon > 0$ be given. Note that $\left|\dfrac{|x|}{x} - (-1)\right| = 0$ if $x < 0$. Hence, let $\delta = 1$ (or any other positive real number); then for every $\epsilon > 0$ there exists $\delta > 0$ $(\delta = 1)$ such that $\left|\dfrac{|x|}{x} - (-1)\right| < \epsilon$ if $x < -\delta$. The domain of the function contains $(\leftarrow, 0)$. Therefore, $\underset{\leftarrow}{lim}\ \dfrac{|x|}{x} = -1$.

5. (a) Ib, Ic, Id, IId, all of Group III.   (b) Ib, Id, IIb, IIc, IId.

7. (a) Ib, Id, IId.   (b) Yes; $\underset{\leftarrow}{lim}\ \dfrac{x-1}{x} = 1,\ \underset{\leftarrow}{lim}\ \dfrac{|x|}{x} = -1,\ \underset{\leftarrow}{lim}\ 2^x = 0$.

2-8 Theorems on Limits

1. Let D_f contain $(b, \rightarrow)$ where b is a real number; then D_f contains $[b + 1, \rightarrow)$. Let D_f contain $[b, \rightarrow)$; then D_f contains $(b, \rightarrow)$.

3. Since $\underset{\rightarrow}{lim} f(x) = -p, p > 0$, and $\frac{1}{2}p > 0$, then by the limit at-the-right definition there exists $\delta > 0$ such that $|f(x) - (-p)| < \frac{1}{2}p$ if $x > \delta$. But, $|f(x) - (-p)| < \frac{1}{2}p$ is equivalent to $-\frac{3}{2}p < f(x) < -\frac{1}{2}p$. Therefore, there exists $\delta > 0$ such that $f(x) < -\frac{1}{2}p$ if $x > \delta$.

5. The domain of the function contains every real number. Since $|c - c| = 0$ and $0 < \epsilon$ for every positive real number ϵ, then for every $\epsilon > 0$ there exists $\delta = 1$ such that $|c - c| < \epsilon$ if $x > \delta$ and $|c - c| < \epsilon$ if $x < -\delta$. Hence, $\underset{\rightarrow}{lim}\ \{(x, c)\} = c,\ \underset{\leftarrow}{lim}\ \{(x, c)\} = c$, and $\underset{\rightarrow}{lim}\ \{(x, c)\} = \underset{\leftarrow}{lim}\ \{(x, c)\} = c$.

7. If a function f has a limit at-the-left L, then f is bounded on some left ray. Proof: If $\underset{\leftarrow}{lim} f(x) = L$, then the domain of f contains an open left ray; and for every $\epsilon > 0$ there exists $\delta > 0$ such that $|f(x) - L| < \epsilon$ if $x < -\delta$. Choose $\epsilon = \frac{1}{2}$ and some $\delta > 0$ guaranteed by the definition so that $|f(x) - L| < \frac{1}{2}$ if $x < -\delta$; that is $L - \frac{1}{2} < f(x) < L + \frac{1}{2}$ if $x < -\delta$. Hence, f is bounded on $(\leftarrow, -\delta)$ by $L - \frac{1}{2}$ and $L + \frac{1}{2}$.

2-9 Another Type of Limit

1. (a) Note that $0 < \dfrac{2}{4n} < \dfrac{1}{n}$ for every natural number n. Since $\lim \dfrac{1}{n} = 0$
then, by the domination principle, $\lim \dfrac{2}{4n} = 0$. (b) Note that $\sin \dfrac{4n\pi}{2} = 0$
for every natural number n. Thus, $\lim \left(\sin \dfrac{4n\pi}{2} \right) = 0$.

3. (a) Yes. (See Exercise 1.) (b) Yes. (See Exercise 2.) (c) No. For every positive
real number d, $\sin x = 0$ for some $x \in (-d, d)$; and $\sin x = 1$ for some other
$x \in (-d, d)$.

5. Yes; for every $\epsilon > 0$, we have found that $|g(x) - 1| < \epsilon$ if $0 < x < \epsilon$ or if
$-\epsilon < x < 0$. This type of limit will be defined formally in 2-11.

2-10 More Examples to Ponder

1. (a) Ia, Ib, IIa, and IIb are monotone increasing; Ic and IIc are not monotone.
(b) All are bounded on $(0, c)$.

3. (a) Ia is monotone increasing; Ib, IIa, and IIb are monotone decreasing; Ic and
IIc are not monotone. (b) All are bounded on $(a, 0)$.

5. (a) There are none.   (b) Ia, 1.

7. (a) $\left| \dfrac{x^2 + x}{x} - 1 \right| = x$ if $x > 0$. Hence, for x in $(0, c)$, $\dfrac{x^2 + x}{x}$ is within $\dfrac{1}{10}$ of
1 if $x < \dfrac{1}{10}$. (b) For x in $(0, c)$, $\dfrac{x^2 + x}{x}$ is within ϵ of 1 if $x < \epsilon$.

2-11 Limits at a Real Number b

1. Ia, 1; Ib, 0; Ic, 0; IIa, 1; IIb, 10^{-6}.

3. Ia, 1; Ib, 0; Ic, 0.

5. Notice that $|c - c| = 0 < \epsilon$ for every $\epsilon > 0$. Hence, let $\delta = 1$ (or any positive
real number); then for every $\epsilon > 0$ there exists $\delta > 0$ ($\delta = 1$) such that $|c - c|$
$< \epsilon$ if $0 < x - b < \delta$. Also, the domain of the function includes $(b, b + 1)$.
Hence, $\lim_{b^+} c = c$. Similarly, $\lim_{b^-} c = c$. Thus, $\lim_{b} c = c$.

7. Let any $\epsilon > 0$ be given. Note that $|mx + b - (ma + b)| = |m||x - a|$ and
$|m||x - a| < \epsilon$ if $|x - a| < \dfrac{\epsilon}{|m|}$ ($m \neq 0$). Hence, choose $\delta = \dfrac{\epsilon}{|m|}$ and it fol-
lows that $|mx + b - (ma + b)| < \epsilon$ if $0 < x - a < \dfrac{\epsilon}{|m|}$ or if $0 < a - x <$
$\dfrac{\epsilon}{|m|}$. The domain of the function contains $\left(-\dfrac{\epsilon}{|m|}, 0 \right)$ and $\left(0, \dfrac{\epsilon}{|m|} \right)$. Hence, $\lim_{a^-} (mx$
$+ b) = ma + b$, $\lim_{a^+} (mx + b) = ma + b$, and consequently $\lim_{a} (mx + b) =$
$ma + b$.

2-12 More Theorems on Limits

1. Assume $L \neq M$. Without loss of generality consider $L > M$; then $\frac{1}{2}(L - M) > 0$. Since $\lim_{b^+} f(x) = L$, then for every $\epsilon > 0$, in particular $\epsilon = \frac{1}{2}(L - M)$, there exists $\delta_1 > 0$ such that $L - \frac{1}{2}(L - M) < f(x) < L + \frac{1}{2}(L - M)$ if $0 < x - b < \delta_1$; that is, $\frac{1}{2}(L + M) < f(x)$ if $0 < x - b < \delta_1$. Since $\lim_{b^+} f(x) = M$, then for every $\epsilon > 0$, in particular $\epsilon = \frac{1}{2}(L - M)$, there exists $\delta_2 > 0$ such that $M - \frac{1}{2}(L - M) < f(x) < M + \frac{1}{2}(L - M)$ if $0 < x - b < \delta_2$; that is, $f(x) < \frac{1}{2}(L + M)$ if $0 < x - b < \delta_2$. Choose $\delta = \min [\delta_1, \delta_2]$. Then for $0 < x - b < \delta$, it follows that $\frac{1}{2}(L + M) < f(x) < \frac{1}{2}(L + M)$; but this is an absurdity. Therefore, our assumption is false, and $L = M$.

3. Since $\lim_{b^+} f(x) = L$, then for $\epsilon = \frac{1}{2}$ there exists $\delta > 0$ such that $|f(x) - L| < \frac{1}{2}$ if $0 < x - b < \delta$; that is, $L - \frac{1}{2} < f(x) < L + \frac{1}{2}$ if $x \in (b, b + \delta)$. Thus, f is bounded by $L - \frac{1}{2}$ and $L + \frac{1}{2}$ on the interval $(b, b + \delta)$.

5. Since $\lim_{b} f(x) = p$ and $p > 0$, then for $\epsilon = \frac{1}{2}p$ there exists $\delta > 0$ such that $|f(x) - p| < \frac{1}{2}p$ if $0 < |x - b| < \delta$. But if $|f(x) - p| < \frac{1}{2}p$, then $f(x) > \frac{1}{2}p > 0$. Hence there exists δ such that $f(x) > \frac{1}{2}p$ if $0 < |x - b| < \delta$.

7. Let any $\epsilon > 0$ be given. Since $\lim_{b} f(x) = L$, there exists $\delta > 0$ such that $L - \epsilon < f(x) < L + \epsilon$ if $0 < |x - b| < \delta$. But if $x \in [(a, b) \cup (b, c)]$, then (i) $g(x) < L + \epsilon$ since $g(x) \leq L$ and (ii) $L - \epsilon < g(x)$ since $f(x) \leq g(x)$. Hence, choose $\delta' = \min [\delta, \min [b - a, c - b]]$, and then it follows that $L - \epsilon < g(x) < L + \epsilon$ if $0 < |x - b| < \delta'$. Therefore, $\lim_{b} g(x) = L$.

2-13 Some Special Limits

1. For every real number x, $0 \leq |x|$ and $|x|^2 = x^2$. Since $|x| < \sqrt{t}$, then $x^2 \leq |x| \sqrt{t}$ (upon multiplying by $|x|$); and $|x| \sqrt{t} < t$ (upon multiplying by $\sqrt{t}$). Thus, since $x^2 \leq |x| \sqrt{t}$ and $|x| \sqrt{t} < t$, then $x^2 < t$.

3. First note that the ray $(\leftarrow, 0)$ is the domain of the function. Now let any $\epsilon > 0$ be given. Since $-(\pi/2) < (\pi/x) < 0$ if $x < -2$, it follows from equation (4) of 2-13 that $(\pi/x) < \sin (\pi/x) < 0$ and $|\sin (\pi/x) - 0| < (\pi/-x)$ if $x < -2$ (keep in mind that $-x > 0$ if $x < -2$). Now to insure that $|\sin (\pi/x) - 0| < \epsilon$, insure that $(\pi/-x) < \epsilon$ and $x < -2$. Thus choose $\delta = \max [2, (\pi/\epsilon)]$, and it follows that if $x < -\delta$, then $x < -2$, $x < -(\pi/\epsilon)$, and $|\sin (\pi/x) - 0| < \epsilon$. Hence, by the limit at-the-left definition, $\lim_{\leftarrow} \sin (\pi/x) = 0$.

5. The domain of the function contains every deleted neighborhood of 0. Let any $\epsilon > 0$ be given. Note that $|1 + \sin x - 1| = |\sin x|$. Hence, following the reasoning in Example 1, we choose $\delta = \min [\epsilon, \pi/4]$; then $|1 + \sin x - 1| = |\sin x| < |x| < \epsilon$ if $0 < |x| < \delta$; $\lim_{0} (1 + \sin x) = 1$.

7. There are none.

2-14 Continuity

1. (i) sin 0 $= 0$; (ii), (iii) $\lim\limits_{0}$ sin $x = 0$. Hence, $\{(x, \sin x)\}$ is continuous at 0.

3. (i) cos 0 $= 1$; (ii), (iii) $\lim\limits_{0}$ cos $x = 1$ by Example 2 of 2-13. Hence, $\{(x, \cos x)\}$ is continuous at 0.

5. Let b be any real number. (i) $b = b$; (ii), (iii) $\lim\limits_{b} x = b$. Hence, $\{(x, x)\}$ is continuous on the set of real numbers.

7. Let b be any real number. (i) $c = c$; (ii), (iii) $\lim\limits_{b} c = c$. Hence $\{(x, c)\}$ is continuous on the set of real numbers.

9. $b = 1$.

11. $b = 1$.

CHAPTER 3—GENERALIZATION AND APPLICATION OF THE LIMIT CONCEPT

3-1 Arithmetic of Functions

1. **(a)** $y = f(x) + g(x)$. **(b)** $y = f(x) - g(x)$. **(c)** $y = f(x)g(x)$. **(d)** $y = f(x)/g(x)$. **(e)** $y = f[g(x)]$ or $y = f(g(x))$.

3. $\left\{\left(x, x + 1 - \dfrac{\sqrt{x}}{|x|}\right)\right\}, x > 0$ 5. $\{(x, |\sqrt{x}|)\}, x \geq 0$.

7. $f \, of \, (g \times g)$ 9. $f \, of \, (h \, of \, g)$.

11. $(f \, of \, g)(x) = x$; $(g \, of \, f)(x) = x$.

13. $(f \, of \, g)(x) = x$; $(g \, of \, f)(x) = x, x > 0$.

15. The four pairs of functions found in Exercises 11 through 14 are examples. (In addition one could name: $f(x) = x + 2$, $g(x) = x - 2$; $f(x) = x + 3$, $g(x) = x - 3$; $f(x) = 4x$, $g(x) = \frac{1}{4}x$; etc.)

3-2 Interrelation of Limit Definitions

1. $(10, \longrightarrow)$, $(11, \longrightarrow)$, $(10^6, \longrightarrow)$.

3. If $\Delta_1 = (10, \longrightarrow)$ and $\Delta_2 = (100, \longrightarrow)$, then the set of real numbers in both Δ_1 and Δ_2 is the set of real numbers in Δ_2.

7. D_f contains $(\longleftarrow, q - 1)$ and in general $(\longleftarrow, -\delta)$ for any real number $-\delta < q$; each of these sets is a δ-set associated with $\lim\limits_{\longleftarrow}$. Hence, D_f contains at least one δ-set.

9. D_f contains $(\frac{1}{2}(a + b), b)$ and in general $(b - \delta, b)$ where $0 < \delta < b - a$; each of these sets is a δ-set associated with $\lim\limits_{b^-}$. Hence, D_f contains at least one δ-set.

13. For every real number $\delta > 0$ there exists some real number x such that $x < -\delta$. Hence, $x \in (\longleftarrow, -\delta)$; thus Δ is non-empty.

15. For every real number $\delta > 0$ there exists some real number x (for example, $x = b - \frac{1}{2}\delta$) such that $b - \delta < x < b$. Hence, $x \in (b - \delta, b)$; thus Δ is nonempty.

3-3 The Generalized Limit

1. (a) For $\underset{\leftarrow}{lim}$, let $\Delta_1 = (\leftarrow, -\delta_1)$ and $\Delta_2 = (\leftarrow, -\delta_2)$. Then $\Delta_3 = \Delta_1 \cap \Delta_2$ $= \{x \mid x < -\delta_3$ where $\delta_3 = \max\,[\delta_1, \delta_2]\} = (\leftarrow, -\delta_3)$. Thus Δ_3 is a δ-set.

(b) For $\underset{b^-}{lim}$, let $\Delta_1 = (b - \delta_1, b)$ and $\Delta_2 = (b - \delta_2, b)$. Then $\Delta_3 = \Delta_1 \cap \Delta_2 = \{x \mid 0 < b - x < \delta_3$ where $\delta_3 = \min\,[\delta_1, \delta_2]\} = (b - \delta_3, b)$. Thus Δ_3 is a δ-set.

(c) For $\underset{b}{lim}$, let $\Delta_1 = [(b - \delta_1, b) \cup (b, b + \delta_1)]$ and $\Delta_2 = [(b - \delta_2, b) \cup (b, b + \delta_2)]$. Then $\Delta_3 = \Delta_1 \cap \Delta_2 = \{x \mid 0 < |x - b| < \delta_3$ where $\delta_3 = \min\,[\delta_1, \delta_2]\} = [(b - \delta_3, b) \cup (b, b + \delta_3)]$. Thus Δ_3 is a δ-set. For each, $\delta_3 \neq \emptyset$.

3. Since $limit\; g(x) = p$ and $p > 0$, we can assert the following: For $\epsilon = \frac{1}{2}p$ there exists a δ-set Δ such that $p - \frac{1}{2}p < g(x) < p + \frac{1}{2}p$ if $x \in \Delta$. Thus, since $p - \frac{1}{2}p = \frac{1}{2}p$ and $\frac{1}{2}p > 0$, there exists a δ-set Δ such that $0 < \frac{1}{2}p < g(x)$ if $x \in \Delta$.

5. Since $limit\; g(x) = M$, it follows from Exercise 3 that there exists a δ-set Δ such that $g(x) > \frac{1}{2}M > 0$ if $x \in \Delta$ and $M > 0$. Similarly, from Exercise 4 it follows that there exists a δ-set Δ such that $g(x) < \frac{1}{2}M < 0$ if $x \in \Delta$ and $M < 0$. If $M > 0$ and $g(x) > \frac{1}{2}M$, then $|g(x)| > \frac{1}{2}|M|$; if $M < 0$ and $g(x) < \frac{1}{2}M$, then $|g(x)| > \frac{1}{2}|M|$. Therefore, for any $M \neq 0$ there exists a δ-set Δ such that $|g(x)| > \frac{1}{2}|M|$ if $x \in \Delta$.

7. Let any $\epsilon > 0$ be given. Since $limit\; f(x) = L$, then there exists a δ-set Δ_1 such that $L - \epsilon < f(x) < L + \epsilon$ if $x \in \Delta_1$. Since $limit\; h(x) = L$, then there exists a δ-set Δ_2 such that $L - \epsilon < h(x) < L + \epsilon$ if $x \in \Delta_2$. We are given that some δ-set Δ exists such that $f(x) \leq g(x) \leq h(x)$ if $x \in \Delta$. Choose $\Delta_3 = [(\Delta_1 \cap \Delta_2) \cap \Delta]$. Then if $x \in \Delta_3$, $L - \epsilon < f(x) \leq g(x) \leq h(x) < L + \epsilon$; that is, there exists a δ-set Δ (determined as Δ_3) such that $|g(x) - L| < \epsilon$ if $x \in \Delta$. Therefore, $limit\; g(x) = L$.

3-4 Generalized Limit Theorems

1. We are given that $limit\, f(x) = L$ and $limit\; g(x) = M$. Let any $\epsilon > 0$ be given. Hence, by the generalized limit definition there exist δ-sets Δ_1 and Δ_2 such that

$$|f(x) - L| < \tfrac{1}{2}\epsilon \quad \text{if} \quad x \in \Delta_1$$

and $$|g(x) - M| < \tfrac{1}{2}\epsilon \quad \text{if} \quad x \in \Delta_2.$$

Select $\Delta_3 = \Delta_1 \cap \Delta_2$. Hence, since $|g(x) - M| = |M - g(x)|$, then $|f(x) - L| + |M - g(x)| < \frac{1}{2}\epsilon + \frac{1}{2}\epsilon = \epsilon$ if $x \in \Delta_3$. But $|(f(x) - g(x)) - (L - M)| \leq |f(x) - L| + |M - g(x)|$, and it follows that

$$|(f(x) - g(x)) - (L - M)| < \epsilon \quad \text{if} \quad x \in \Delta_3.$$

Thus, since ϵ is any positive real number we can conclude by the generalized limit definition that $limit\; (f(x) - g(x)) = L - M$.

3. (b) $Lim_b\, 2x^3 = 2b^3$, $\lim_b 3x = 3b$ (Theorem 3–4c and Example 3 of 3–4) and $\lim_b 7 = 7$. Hence, by Theorems 3–4a and 3–4b $\lim_b (2x^3 - 3x + 7) = 2b^3 - 3b + 7$. **(c)** $Lim_b\, x^2 = b^2$ and $\lim_b 1 = 1$. Hence, $\lim_b (x^2 - 1) = b^2 - 1$ and $\lim_b (x^2 + 1) = b^2 + 1$. Since $b^2 + 1 \neq 0$, then by Theorem 3–4e $\lim_b \dfrac{x^2 - 1}{x^2 + 1} = \dfrac{b^2 - 1}{b^2 + 1}$. **(d)** $Lim_0\, 3 = 3$; and since $\lim_0 \dfrac{\sin x}{x} = 1$, then $\lim_0 \dfrac{\sin^2 x}{x^2} = 1 \cdot 1 = 1$ by Theorem 3–4d. Therefore, $\lim_0 \dfrac{3 \sin^2 x}{x^2} = 3 \cdot 1 = 3$ by Theorem 3–4c.

5. (a) $Lim_0\, x = 0$, and $|\sin (1/x)| \leq 1$ for $x \in [(-\pi, 0) \cup (0, \pi)]$. Hence, $\lim_0 \left(x \sin \dfrac{1}{x} \right) = 0$. **(b)** $Lim_{\rightarrow} \dfrac{1}{x} = 0$, and $|\sin x| \leq 1$ for $x \in (0, \rightarrow)$. Hence, $\lim_{\rightarrow} \left(\dfrac{\sin x}{x} \right) = 0$. **(c)** $Lim_5 (x - 5) = 5 - 5 = 0$, and $|\cos (1/x)| \leq 1$ for x in every deleted neighborhood of 5. Hence, $\lim_5 \left((x - 5) \cos \dfrac{1}{x} \right) = 0$. **(d)** $Lim_{\leftarrow} 2^x = 0$ (Example 2 of 2–7); and $\dfrac{|x|}{x} = -1$ for $x < 0$. Hence, $\lim_{\leftarrow} \dfrac{|x| 2^x}{x} = 0$.

3-5 More on Composition and Continuity

1. The function is a polynomial. Hence it is continuous at every real number b by Theorem 3–5a.

3. The functions whose equations are $y = 3$, $y = 5$, $y = 7$, and $y = x$ are continuous at any real number. Hence, by Theorem 3–5b it follows that $y = 3/x^2$, $y = 5/x$, and consequently $y = (3/x^2) - (5/x) + 7$ are equations of functions that are continuous at any real number b except 0.

5. Note that $(1 + x)/x = (1/x) + 1$, $\lim_{\rightarrow} (1/x) = 0$, and $\lim_{\rightarrow} 1 = 1$. Hence, $\lim_{\rightarrow} ((1 + x)/x) = 1$. Since f is continuous at 1, $\lim_{\rightarrow} \sqrt{(1 + x)/x} = \sqrt{1} = 1$.

7. The function $g = \{(x, x - b)\}$ is continuous at b, $g(b) = b - b = 0$, and the sine function is continuous at 0 (Exercise 1 of 2–14). Hence, by Theorem 3-5d, F is continuous at b.

9. $Lim_b \sin x = \lim_b [\sin b \cos(x - b) + \cos b \sin (x - b)]$; but since $\{(x, \cos (x - b))\}$ and $\{(x, \sin (x - b))\}$ are continuous at b (Exercises 7 and 8), then $\lim_b \cos (x - b) = 1$ and $\lim_b \sin (x - b) = 0$. Hence, $\lim_b \sin x = (\sin b)(1) + (\cos b)(0) = \sin b$. But since $\lim_b \sin x = \sin b$, the sine function is continuous at b for every real number b.

3-6 Limits in High School Mathematics

1. (a) $s_n - rs_n = a - ar^n$.   **(b)** $s_n = \dfrac{a(1 - r^n)}{1 - r}$.

3. Assume that $\lim s_n$ does exist; that is, assume there is some real number L such that $lim\, s_n = L$. However, if $s_n = \dfrac{a(1 - r^n)}{1 - r}$, then $r^n = 1 - \dfrac{(s_n - rs_n)}{a}$. By

proved limit theorems, $lim\, r^n = lim\left(1 - \dfrac{s_n - rs_n}{a}\right) = 1 - \dfrac{L - rL}{a}$. But this is

impossible since $lim\, r^n$ was given not to exist; hence, our assumption is false, and $lim\, s_n$ does not exist.

5. Note that $\dfrac{\pi(r + h)^2 - \pi r^2 h}{h} = 2\pi r + \pi h$. Hence,

$$\lim_{0^+} \frac{\pi(r + h)^2 - \pi r^2 h}{h} = \lim_{0^+}(2\pi r + \pi h) = 2\pi r$$

because $\lim_{0^+} h = 0$ and π and r are constants for the given circle.

3-7 On from Here . . .

1. Let any $M > 0$ be given. Then $2x > M$ if $x > \frac{1}{2}M$. Hence, let $\delta = \frac{1}{2}M$, and then $f(x) > M$ if $x > \delta$; f increases without bound at-the-right.

3. The point A.

Answers to
Even-Numbered Exercises

CHAPTER 1—SEQUENCES AND THEIR LIMITS

1-1 Infinite Sequences

2. $1, \frac{1}{2}, \frac{1}{3}, \frac{1}{4}, \frac{1}{5}$.

4. $\{(n, 1)\}$.

6. $c, c + d, c + 2d, c + 3d, c + 4d$.

8. 1, 1, 2, 3, 5, 8, 13.

1-2 Graphs of Infinite Sequences

2.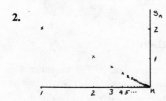

4.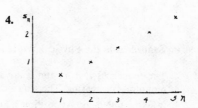

6.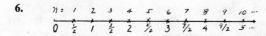

1-3 Some Examples to Ponder

2. (a) through (c) IIb, IIIc, IIId.

4. (a) $N = \frac{1}{2}$, $M = -1$ (N and M could be other numbers providing $N \geq \frac{1}{2}$, $M \leq -1$). (b) $N = 1$, $M = 0$ (N and M could be other numbers providing $N \geq 1$, $M \leq 0$). (c) $N = 1$, $M = -1$ (N and M could be other numbers providing $N \geq 1$, $M \leq -1$).

6. Id, IIb.

1-4 Types of Sequences

2. The sequences in Group I are bounded. The sequences in Group II are not bounded, but IIa, IIc, and IId are bounded below while IIb is bounded above. IIIa and IIIb are bounded, but IIIc and IIId are not bounded above or below.

4. The least upper bounds are as follows: for Ia, $\frac{1}{2}$; for Ib, 1; for Ic, $\frac{1}{4}$; for Id, 2; for IIb, -1; for IIIa and IIIb, 1.

6. The greatest lower bounds are as follows: for Ia, -1; for Ib, 0; for Ic, $-\frac{1}{2}$; for Id, IIa, and IIc, 1; for IId, -9; for IIIa and IIIb, -1.

8. Given a sequence s and two greatest lower bounds of s, say m and m', we must prove that $m = m'$. If $m < m'$, then m would not be a greatest lower bound; and likewise, if $m' < m$, then m' would not be a greatest lower bound. Hence, it must be that $m = m'$; any two greatest lower bounds of a sequence s are equal.

1-5 Circumference of a Circle

2. (a) Bounded (above by C, below by $2\sqrt{2}$). (b) Monotone increasing.

4. The least upper bound of $\{(n, p_n)\}$ is π.

6. π.

1-6 Sequences in the Physical World and in History

2. The sequences t, a, r, and b are monotone increasing. The sequence h is monotone decreasing. The sequence d is not monotone.

4. (a) No. (b) No. (c) At position 2. (d) At position 2. (e) Yes, providing one or more seconds are permitted to elapse.

6. (a) No. (b) At the point where displacement is 0. This example is important: the pendulum "reaches" its rest position infinitely many times, and the rest position 0 is neither the least upper bound nor greatest lower bound of the sequence d.

1-7 The Limit of a Sequence: Informal

2. $n > k$.

4. The sets of natural numbers in the answers to the Example, Exercise 1, and Exercise 3 are a finite set $\{n \mid 34 \leq n \leq 99\}$, an infinite set $\{n \mid n > 100\}$ (in fact, all natural numbers except a finite set), and the empty set, respectively.

6. Recall that $(n - 1)/n = 1 - \dfrac{1}{n}$. Then $1 - \dfrac{1}{n} > 0.9 - 1/100$ if $n > \dfrac{100}{11}$ and $1 - \dfrac{1}{n} < 0.9 + 1/100$ if $n < \dfrac{100}{9}$; that is, $(n - 1)/n$ is within $1/100$ of 0.9 if $\dfrac{100}{11} < n < \dfrac{100}{9}$.

8. The sets of natural numbers in the answers to Exercises 5, 6, and 7 are an infinite set $\{n \mid n > k\}$, a finite set $\{10, 11\}$, and the empty set, respectively.

1-8 A Precise Language

2. $n > \dfrac{1}{\epsilon}$.   **4.** $n > \dfrac{1}{\epsilon}$.

6. (a) If $a \geq 0$, then $-a \leq 0$. Hence, $|a| = a$ and $|-a| = a$. If $a < 0$, then $-a > 0$. Hence, $|a| = -a$ and $|-a| = -a$. Therefore, in either case, $|a| = |-a|$. (b) $|ab| = ab = |a||b|$ if $a = 0$ or $b = 0$, if $a > 0$ and $b > 0$, or if $a < 0$ and $b < 0$; $|ab| = -ab = |a||b|$ if $a > 0$ and $b < 0$, or if $a < 0$ and $b > 0$. (c) $|a + b| = |a| + |b|$ if a and b have like signs or if either a or b or both are 0; $|a + b| = \|a\| - |b\| < |a| + |b|$ if $a > 0$ and $b < 0$; $|a + b| = |-|a| + |b|| < |a| + |b|$ if $a < 0$ and $b > 0$. (d) $|s_n - L| < \epsilon$ if (i) $s_n - L \geq 0$ and $s_n - L < \epsilon$ or (ii) $s_n - L < 0$ and $-(s_n - L) < \epsilon$. The conditions of (i) yield $L \leq s_n < L + \epsilon$; the conditions of (ii) yield $L - \epsilon < s_n < L$. Both (i) and (ii) hold iff $L - \epsilon < s_n < L + \epsilon$. Therefore, $|s_n - L| < \epsilon$ iff $L - \epsilon < s_n < L + \epsilon$.

1-9 Limit of a Sequence

2. Let $\delta = \dfrac{1}{\epsilon}$; then for every $\epsilon > 0$ there exists $\delta > 0$ $\left(\delta = \dfrac{1}{\epsilon}\right)$ such that $\left|\dfrac{n - 1}{n} - 1\right| < \epsilon$ if $n > \delta$. Therefore, $\lim \dfrac{n - 1}{n} = 1$.

4. We shall prove that for some $\epsilon > 0$ $\left(\text{specifically } \epsilon = \dfrac{1}{200}\right)$ there is no δ such that $\left|\dfrac{1}{n} - \dfrac{1}{100}\right| < \epsilon$ if $n > \delta$. We use the indirect approach; assume there is some $\delta > 0$ such that $\left|\dfrac{1}{n} - \dfrac{1}{100}\right| < \dfrac{1}{200}$ if $n > \delta$. Now, we select a natural number n such that $n > \delta$ and $n > 400$. Since $n > 400$, $\dfrac{1}{n} < \dfrac{1}{400}$ and $\left|\dfrac{1}{n} - \dfrac{1}{100}\right| > \dfrac{3}{400}$; but $\dfrac{3}{400} > \dfrac{1}{200}$. Thus, our assumption has lead to a contradiction, and we conclude that for some $\epsilon > 0$ there is no $\delta > 0$ that satisfies the conditions of the limit definition. Therefore, $\lim \dfrac{1}{n} \neq \dfrac{1}{100}$.

6. First note that $2^1 > 1$. Assume that $2^k > k$ for some positive integer k. Then since $2(2^k) > 2k$, $2(2^k) = 2^{k+1}$, and $2k = k + k \geq k + 1$, it follows that $2^k > k$ implies $2^{k+1} > k + 1$. By the principle of mathematical induction, $2^n > n$ for every natural number n. Therefore, $\dfrac{1}{2^n} < \dfrac{1}{n}$ for every natural number n.

1-10 Graphical Interpretation of the Limit

2. No. When n is odd and $n > 3$, each point (n, s_n) is below the line $y = -\frac{1}{2}$. When n is even and $n \geq 4$, each point (n, s_n) is above the line $y = \frac{1}{2}$. When $\epsilon = \frac{1}{2}$, no ϵ-band includes points that are below $y = -\frac{1}{2}$ and above $y = \frac{1}{2}$.

4. (a) Of the many correct answers are: $\{(n, (-1)^n 2)\}$, and $\left\{\left(n, (-1)^n \dfrac{2n-1}{n}\right)\right\}$.

(b) Of the many correct answers are: $\{(n, s_n)\}$ where $s_n = \dfrac{1}{n}$ if $n = 3k$, $s_n = 1$ if $n = 3k - 2$, $s_n = 2$ if $n = 2k - 1$ where k is a natural number.

6. No. Assume L and M are both cluster points and $L > M$. Let $\epsilon = \dfrac{L-M}{2}$. If some real number X is to be $\lim s_n$, there must exist some real number $\delta > 0$ such that $|s_n - X| < \dfrac{L-M}{2}$ for every $n > \delta$. However, for every $\delta > 0$ there exist natural numbers m and m' such that $|s_m - L| < \dfrac{L-M}{2}$ and $|s_{m'} - M| < \dfrac{L-M}{2}$. Thus, s_m and $s_{m'}$ satisfy the inequalities $s_m > \dfrac{L+M}{2}$ and $s_{m'} < \dfrac{L+M}{2}$, and s_m and $s_{m'}$ are not within $\dfrac{L-M}{2}$ of any real number X. Therefore, $\lim s_n$ does not exist.

1-11 Theorems on Limits

2. Let any $\epsilon > 0$ be given and let $\delta = 1$. Since $c_n = c$, it follows that $|c_n - c| = |c - c| = 0$ for every n. Hence, $|c_n - c| < \epsilon$ if $n > \delta$. Therefore, by the limit definition, $\lim c_n = c$.

4. Given $\lim s_n = L$. Consider $\{(n, u_n)\} = s_{1+j}, s_{2+j}, \ldots, s_{n+j}, \ldots$ where s_1 through s_j have been omitted. Prove that $\{(n, u_n)\}$ has a limit. The proof that $\lim u_n = L$ can be obtained from the proof of Exercise 3 by changing "4" to "j."

6. Given $\lim s_n = L$. Consider $\{(n, t_n)\} = x_1, x_2, \ldots, x_j, s_1, s_2, \ldots, s_{n-j}, \ldots$ where the x's are real numbers. Prove that $\{(n, t_n)\}$ has a limit. The proof that $\lim t_n = L$ can be obtained from the proof of Exercise 5 by changing "3" to "j."

1-12 Applications of Limits of Sequences

2. $0.9999\cdots$.

4. Notice that $\left| t_n - \dfrac{1}{3} \right| = \left| \dfrac{333\cdots3}{10^n} - \dfrac{1}{3} \right| = \left| -\dfrac{1}{3\cdot10^n} \right| = \dfrac{1}{3\cdot10^n}$. But $\dfrac{1}{3\cdot10^n} < \dfrac{1}{10^n} < \dfrac{1}{n} < \epsilon$ if $n > \dfrac{1}{\epsilon}$. Hence, $\left| t_n - \dfrac{1}{3} \right| < \epsilon$ if $n > \dfrac{1}{\epsilon}$. Therefore, for every $\epsilon > 0$ there exists $\delta > 0$ $\left(\delta = \dfrac{1}{\epsilon} \right)$ such that $\left| t_n - \dfrac{1}{3} \right| < \epsilon$ if $n > \delta$. Thus, $\lim t_n = \dfrac{1}{3}$.

6. For every natural number n, $r_n + \dfrac{1}{10^n} > \sqrt{2}$.

8. Since $|r_n - \sqrt{2}| < \dfrac{1}{10^n}, \dfrac{1}{10^n} < \dfrac{1}{n}$, and $\delta = \dfrac{1}{\epsilon}$, then if $n > \delta$ we have the following conclusions: $n > \dfrac{1}{\epsilon}$, and $|r_n - \sqrt{2}| < \dfrac{1}{10^n} < \dfrac{1}{n} < \epsilon$. Thus for every $\epsilon > 0$ there exists $\delta > 0$ $\left(\delta = \dfrac{1}{\epsilon} \right)$ such that $|r_n - \sqrt{2}| < \epsilon$ if $n > \delta$. Thus, $\lim r_n = \sqrt{2}$.

CHAPTER 2—FUNCTIONS AND THEIR LIMITS

2-1 Functions

2. (a) $G(-3) = 3(-3) = -9$. (b) $G(3) = 3(3) = 9$. (c) $G(5 - 7) = 3(5 - 7) = -6$. (d) $G(a + b) = 3(a + b) = 3a + 3b$. (e) $G(a) + G(b) = 3a + 3b$. (f) $G(a\cdot b) = 3(a\cdot b) = 3ab$. (g) $G(a)\cdot G(b) = 3a\cdot3b = 9ab$.

4. $y = 5$ or $f(x) = 5$.

2-2 Graphs of Functions

2. $x = n\pi$, where $n = 1, 2, 3, \dots$.

4.

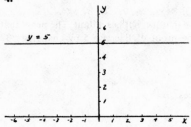

6. The graphs of $\{(n, 5)\}$ and $\{(n, n)\}$ are discrete sets of points (they are a fixed distance apart) on the graphs of $\{(x, 5)\}$ and $\{(x, x)\}$, respectively.

8. (a) $y = -\sqrt{1 - x^2}$

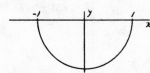

(b) The graph of g is the "bottom half" of a circle whose top half is the graph of the function f in Figure 4.

2-3 Some Examples to Ponder

2. All of Group I, IIa, IIIa, IIIc, IIId.

4. All of Group I, IIb, IIc, IId, all of Group III.

6. Ia, Ic.

8. (a) Ic, Id, all of Group III. **(b)** Ia and Ib are excluded from the answers to part (a) but are included in the answers to Exercise 7.

10. Ia, Id, IIa.

2-4 Classification of Functions

2. (a) All of Group I, IIa, all of Group III. **(b)** Ib, Id, IIb, IIc, IId.

4. All of Group I, IIb, IIc, IId, all of Group III. **(b)** Ia, Id, IIa.

6. If f is bounded on a subset S of the domain of f, then there exists real numbers $\overline{M}$ and N such that $\overline{M} < f(x) < N$ for every x in S. If $|N| \leq |\overline{M}|$, choose $M = |\overline{M}|$; if $|N| > |\overline{M}|$ choose $M = |N|$. In either case, $N \leq M$ and $-M \leq \overline{M}$. Therefore, $-M < f(x) < M$ for every x in S; that is, $|f(x)| < M$

for every x in S. Conversely, if M is a positive real number and $|f(x)| < M$ for every x in S, then $-M < f(x) < M$. Hence, f is bounded above by M and below by $-M$ on S. Therefore, f is bounded on S iff $|f(x)| < M$.

2-5 A Limit of a Function: Informal

2. (a) $x > \dfrac{\pi}{2}$. (b) $x > \dfrac{5\pi}{2}$. (c) Yes. (This is a conclusion based on intuition—a definition will be given in the next section.)

4. $x > \dfrac{1}{\epsilon}$.   **6.** $x > 0$.

2-6 The Limit At-The-Right

2. Let any $\epsilon > 0$ be given. Note that $\left|\dfrac{x-1}{x} - 1\right| = \dfrac{1}{x}$ if $x > 0$ and $\dfrac{1}{x} < \epsilon$ if $x > \dfrac{1}{\epsilon}$. Hence, let $\delta = \dfrac{1}{\epsilon}$; then for every $\epsilon > 0$ there exists $\delta > 0$ $\left(\delta = \dfrac{1}{\epsilon}\right)$ such that $\left|\dfrac{x-1}{x} - 1\right| < \epsilon$ if $x > \delta$. The domain of the function contains $(0, \rightarrow)$. Therefore, $\underset{\rightarrow}{lim} \dfrac{x-1}{x} = 1$.

4. We shall prove that $\underset{\rightarrow}{lim} \dfrac{1}{x-1} = 0$. Notice that $\left|\dfrac{1}{x-1} - 0\right| < \epsilon$ if $\dfrac{1}{\epsilon} < x - 1$ and $\dfrac{1}{x-1} > 0$; that is, if $x > \dfrac{1}{\epsilon} + 1$. Let $\delta = \dfrac{1}{\epsilon} + 1$; then for every $\epsilon > 0$ there exists $\delta > 0$ $\left(\delta = \dfrac{1}{\epsilon} + 1\right)$ such that $\left|\dfrac{1}{x-1} - 0\right| < \epsilon$ if $x > \delta$. The domain of the function contains $(1, \rightarrow)$. Therefore, $\underset{\rightarrow}{lim} \dfrac{1}{x-1} = 0$.

6. Since $\left|\dfrac{\sin x}{x} - 0\right| \le \left|\dfrac{1}{x}\right| = \dfrac{1}{x}$ if $x > 0$, let $\delta = \dfrac{1}{\epsilon}$. Then for every $\epsilon > 0$ there exists $\delta > 0$ $\left(\delta = \dfrac{1}{\epsilon}\right)$ such that $\left|\dfrac{\sin x}{x} - 0\right| < \epsilon$ if $x > \delta$. The domain of the function contains $(0, \rightarrow)$. Hence, $\underset{\rightarrow}{lim} \dfrac{\sin x}{x} = 0$.

8. (a) $\{1, -1\}$. (b) $\underset{\rightarrow}{lim} \dfrac{|x|}{x} = 1$, $\underset{\rightarrow}{lim}\left(-\dfrac{|x|}{x}\right) = -1$.

2-7 The Limit At-The-Left

2. Let any $\epsilon > 0$ be given. Note that $\left|\dfrac{x-1}{x} - 1\right| = -\dfrac{1}{x}$ if $x < 0$ and $-\dfrac{1}{x} < \epsilon$

if $x < -\dfrac{1}{\epsilon}$. Hence, let $\delta = \dfrac{1}{\epsilon}$; then for every $\epsilon > 0$ there exists $\delta > 0$ $\left(\delta = \dfrac{1}{\epsilon}\right)$

such that $\left|\dfrac{x-1}{x} - 1\right| < \epsilon$ if $x < -\delta$. The domain of the function contains

$(\leftarrow, 0)$. Therefore, $\underset{\leftarrow}{lim}\ \dfrac{x-1}{x} = 1$.

4. Ia, Ic, Id, IIc, IId, all of Group III. **(b)** Ia, Id, IIa.

6. (a) Ia, Id. **(b)** Yes; $\underset{\leftarrow}{lim}\ \dfrac{1}{x} = 0$, $\underset{\leftarrow}{lim}\ \dfrac{|x|}{x} = -1$.

2-8 Theorems on Limits

2. Given $|f(x) - L| < \epsilon$ for every $x > \delta$, and let $\delta' = \delta + 1$. Hence, $|f(x) - L|$ $< \epsilon$ for every $x \geq \delta'$ because $\delta' > \delta$. Given $|f(x) - L| < \epsilon$ for every $x \geq \delta'$, and let $\delta = \delta'$. Hence, $|f(x) - L| < \epsilon$ for every $x > \delta$ because $x \geq \delta'$ if $x > \delta$.

4. Since $\underset{\leftarrow}{lim}\ f(x) = p$, $p > 0$, and $\dfrac{p}{2} > 0$, then by the limit at-the-left definition,

there exists $\delta > 0$ such that $|f(x) - p| < \dfrac{p}{2}$ if $x < -\delta$. But, $|f(x) - p| < \dfrac{p}{2}$

is equivalent to $\dfrac{p}{2} < f(x) < \dfrac{3p}{2}$. Therefore, there exists $\delta > 0$ such that

$f(x) > \dfrac{p}{2}$ if $x < -\delta$.

6. Since $\underset{\rightarrow}{lim}\ f(x) = L$, for every $\epsilon > 0$ there exists $\delta > 0$ such that $L - \epsilon <$ $f(x) < L + \epsilon$ if $x > \delta$. But since $L \leq g(x)$, then $L - \epsilon < g(x)$. Note that $g(x) \leq f(x)$ if $x > a$ and $f(x) < L + \epsilon$ if $x > \delta$. Thus, choose $\delta' = \delta$ if $\delta \geq a$ or $\delta' = a$ if $a > \delta$; then, $g(x) < L + \epsilon$ if $x > \delta'$. Hence, for every $\epsilon > 0$ there exists $\delta' > 0$ such that $L - \epsilon < g(x) < L + \epsilon$ if $x > \delta'$. We are given that D_g contains $(a, \rightarrow)$. Therefore, $\underset{\rightarrow}{lim}\ g(x) = L$.

2-9 Another Type of Limit

2. (a) Note that $0 < \dfrac{2}{4n+1} < \dfrac{1}{2n}$ for every natural number n, and $lim\ \dfrac{1}{2n} = 0$.

Therefore, by the domination principle, $lim\ \dfrac{2}{4n+1} = 0$. **(b)** Since

$\sin\dfrac{(4n+1)\pi}{2} = 1$ for every natural number n, $lim\left(\sin\dfrac{(4n+1)\pi}{2}\right) = 1$.

4. Analysis of f:

On $(\leftarrow, 0)$, f is monotone increasing and bounded below by 1.
If $x < -10 < 0$, then

$$\left|\left(\frac{x-1}{x}\right) - 1\right| = \left|-\frac{1}{x}\right| < \frac{1}{10}.$$

If $x < -100 < 0$, then

$$\left|\left(\frac{x-1}{x}\right) - 1\right| = \left|-\frac{1}{x}\right| < \frac{1}{100}.$$

Let any $\epsilon > 0$ be given.
Then $|f(x) - 1| < \epsilon$ if

$$x < -\frac{1}{\epsilon} < 0; \; \lim_{\leftarrow} f(x) = 1.$$

Analysis of g:

On $(\leftarrow, 0)$, g is monotone decreasing and bounded below by 1.
If $-\frac{1}{10} < x < 0$, then $|(1-x) - 1| =$

$$|-x| < \frac{1}{10}.$$

If $-\frac{1}{100} < x < 0$, then

$$|(1-x) - 1| = |-x| < \frac{1}{100}.$$

Let any $\epsilon > 0$ be given.
Then $|g(x) - 1| < \epsilon$ if
$-\epsilon < x < 0$; limit ?????

2-10 More Examples to Ponder

2. No difference.

4. No difference.

6. (a) Ia, 1; Ib, 0; IIa, 1; IIb, $\frac{1}{10^6}$. (b) Ib, 0; IIa, -1; IIb, $-\frac{1}{10^6}$.

8. (a) $\left|\frac{|x|}{x}(x+1) - 1\right| = x$ if $x > 0$. Hence for x in $(0, c)$, $\frac{|x|}{x}(x+1)$ is within $\frac{1}{10}$ of 1 if $x < \frac{1}{10}$. (b) For x in $(0, c)$, $\frac{|x|}{x}(x+1)$ is within ϵ of 1 if $x < \epsilon$.

2-11 Limits at a Real Number b

2. Ia, 1; Ib, 0; Ic, 0; IIa, -1; IIb, $-\frac{1}{10^6}$.

4. Let any $\epsilon > 0$ be given. Notice that $|\sqrt{x} - 0| = \sqrt{x}$ and $\sqrt{x} < \epsilon$ if $0 < x < \epsilon^2$. Hence, choose $\delta = \epsilon^2$, and then for every $\epsilon > 0$ there exists $\delta > 0 \, (\delta = \epsilon^2)$ such that $|\sqrt{x} - 0| < \epsilon$ if $0 < x - 0 < \delta$. Also, D_f includes the interval $(0, \epsilon^2)$ for every $\epsilon > 0$. Hence $\lim_{0^+} \sqrt{x} = 0$.

6. Notice that $\left|x \sin\frac{1}{x} - 0\right| = \left|x \sin\frac{1}{x}\right| \leq |x|$ for every $x \neq 0$ and $|x| < \epsilon$ if $0 < x - 0 < \epsilon$ or if $0 < 0 - x < \epsilon$. Thus, choose $\delta = \epsilon$, and then for every $\epsilon > 0$ there exists $\delta > 0 \, (\delta = \epsilon)$ such that $\left|x \sin\frac{1}{x} - 0\right| < \epsilon$ if $0 < x - 0 < \delta$

and $\left| x \sin \dfrac{1}{x} - 0 \right|$ if $0 < 0 - x < \delta$. Also, D_H includes the intervals $(0, \delta)$ and $(-\delta, 0)$ for every positive value of δ. Hence, $\displaystyle\lim_{0+} \left(x \sin \dfrac{1}{x} \right) = 0$, $\displaystyle\lim_{0-} \left(x \sin \dfrac{1}{x} \right) = 0$, and $\displaystyle\lim_{0} \left(x \sin \dfrac{1}{x} \right) = 0$.

8. For all real numbers a and b where $a < b$ and $b < c$, the intervals (a, b) and (b, c) are in the domain of the function. Let any $\epsilon > 0$ be given. If x is any real number, then $|x^2 - b^2| = |x + b||x - b|$. Because we want δ-conditions on $(x - b)$ and $(b - x)$ let us choose $a = b - 1$ and $c = b + 1$, so that $2b - 1 < x + b < 2b + 1$ if $x \in (b - 1, b)$ or if $x \in (b, b + 1)$; that is, if $-1 < x - b < 1$. Here we note that if $-1 < x - b < 1$, then $|x + b| <$ max $[|2b - 1|, |2b + 1|]$ which we call m. Now, if $-1 < x - b < 1$, then $|x + b||x - b| < m|x - b|$; and if $|x - b| < \epsilon/m$, then $m|x - b| < \epsilon$. Hence, choose $\delta_l = \delta_r = \min [\epsilon/m, 1]$, and then $|x + b||x - b| < \epsilon$ if $0 < x - b < \delta_r$ or if $0 < b - x < \delta_l$. Therefore, for every $\epsilon > 0$ there exists δ_l and δ_r $(\delta_l = \delta_r = \min [\epsilon/m, 1])$ such that $|x^2 - b^2| < \epsilon$ if $0 < x - b < \delta_r$ and $|x^2 - b^2| < \epsilon$ if $0 < b - x < \delta_l$; that is, $\displaystyle\lim_{b+} x^2 = b^2$, $\displaystyle\lim_{b-} x^2 = b^2$ and thus $\displaystyle\lim_{b} x^2 = b^2$.

2-12 More Theorems on Limits

2. Assume $L \neq M$. Without loss of generality consider $L > M$; then $\dfrac{L - M}{2} > 0$. Since $\displaystyle\lim_{b-} f(x) = L$ and $\displaystyle\lim_{b-} f(x) = M$, then for every $\epsilon > 0$, in particular $\epsilon = \dfrac{L - M}{2}$, there exists $\delta_1 > 0$ and $\delta_2 > 0$ such that $L - \dfrac{L - M}{2} < f(x) < L + \dfrac{L - M}{2}$ if $0 < b - x < \delta_1$ and $M - \dfrac{L - M}{2} < f(x) < M + \dfrac{L - M}{2}$ if $0 < b - x < \delta_2$. Choose $\delta = \min [\delta_1, \delta_2]$ and it follows that $\dfrac{L + M}{2} = L - \dfrac{L - M}{2} < f(x)$ and $f(x) < M + \dfrac{L - M}{2} = \dfrac{L + M}{2}$ if $0 < b - x < \delta$. But $\dfrac{L + M}{2} < f(x) < \dfrac{L + M}{2}$ is an absurdity. Hence, our assumption that $L \neq M$ is false, and $L = M$.

4. Since $\displaystyle\lim_{b-} f(x) = L$, then for every $\epsilon > 0$ there exists $\delta_1 > 0$ such that $|f(x) - L| < \epsilon$ if $0 < b - x < \delta_1$; that is, if $x \in (b - \delta_1, b)$. Also, $f(x) = g(x)$ for $x \in (a, b)$. Select $\delta_2 = \min [\delta_1, b - a]$. Then when $x \in (b - \delta_2, b)$ it follows that $f(x) = g(x)$ and $|f(x) - L| < \epsilon$ and, consequently, $|g(x) - L| < \epsilon$. Therefore, for every $\epsilon > 0$ there exists $\delta > 0$ (determined as δ_2) such that $|g(x) - L| < \epsilon$ if $x \in (b - \delta, b)$; that is, $\displaystyle\lim_{b-} g(x) = L$.

6. Let any $\epsilon > 0$ be given. Notice that $|x - b| < \epsilon$ if $0 < |x - b| < \epsilon$. Hence, choose $\delta = \epsilon$; then for every $\epsilon > 0$ there exists $\delta > 0$ $(\delta = \epsilon)$ such that $|x - b| < \epsilon$ if $0 < |x - b| < \delta$. The domain of the function contains every deleted neighborhood of b (it even contains b). Therefore, $\displaystyle\lim_{b} x = b$.

2-13 Some Special Limits

2. Note that $\left| 1 - \dfrac{x^2}{2} - 1 \right| = \dfrac{x^2}{2}$, and $\dfrac{x^2}{2} < \epsilon$ if $x^2 < 2\epsilon$ or if $0 < x < \sqrt{2\epsilon}$.

Hence, choose $\delta = \sqrt{2\epsilon}$ and it follows from the definition that $\lim\limits_{0} \left(1 - \dfrac{x^2}{2} \right) = 1$.

4. Since $b > 0$, it follows that $\left| \dfrac{|x|}{x} - 1 \right| = 0 < \epsilon$ for every $\epsilon > 0$ if $x \in \left(b - \dfrac{b}{2}, b + \dfrac{b}{2} \right)$. Hence, choose $\delta = \dfrac{b}{2}$ and it follows from the definition that $\lim\limits_{b} \dfrac{|x|}{x} = 1$ if $b > 0$.

6. If $\epsilon < 1$, we use the hint and choose $\delta_l = \dfrac{\log (1 - \epsilon)}{\log 2}$ and $\delta_r = \dfrac{\log (1 + \epsilon)}{\log 2}$. If $\epsilon \geq 1$, we arbitrarily select 0.5 so that $0.5 < 1 < \epsilon$; and we choose $\delta_l = \dfrac{\log 0.5}{\log 2}$ and $\delta_r = \dfrac{\log 1.5}{\log 2}$. Then it follows from the definition that $\lim\limits_{0} 2^x = 1$.

2-14 Continuity

2. (i) $1 + \sin 0 = 1$; (ii), (iii) $\lim\limits_{0} (1 + \sin x) = 1$. Hence, $\{(x, 1 + \sin x)\}$ is continuous at 0.

4. (i) $0^2 = 0$; (ii), (iii) $\lim\limits_{0} x^2 = 0^2$ by Exercise 8 of 2-11. Hence, $\{(x, x^2)\}$ is continuous at 0.

6. Let b be any real number. (i) $x^2 = b^2$ when $x = b$; (ii), (iii) $\lim\limits_{b} x^2 = b^2$. Hence, $\{(x, x^2)\}$ is continuous on the set of real numbers.

8. Let a be any real number. (i) $mx + b = ma + b$ when $x = a$; (ii), (iii) $\lim\limits_{a} (mx + b) = ma + b$ by Exercise 7 of 2-11. Hence, $\{(x, mx + b)\}$ is continuous on the set of real numbers.

10. $b = 0$.   **12.** There is none.

CHAPTER 3—GENERALIZATION AND APPLICATION OF THE LIMIT CONCEPT

3-1 Arithmetic of Functions

2. $\{(x, x^2 + x)\}$, $x \geq 0$.   **4.** $\{(x, \sqrt{|x|})\}$.

6. $g \times g$.   **8.** f of f.

10. (a) f of g is g. (b) g of f is g.

12. $(f \text{ of } g)(x) = x$, $x \geq 0$; $(g \text{ of } f)(x) = x$, $x \geq 0$.

14. $(f \text{ of } g)(x) = x$, $x \neq 0$; $(g \text{ of } f)(x) = x$, $x \neq 0$.

3-2 Interrelation of Limit Definitions

2. $(100, \rightarrow)$, $(101, \rightarrow)$, $(10^6, \rightarrow)$.

4. Let δ be a positive real number. **(a)** $(2 - \delta, 2) \cup (2, 2 + \delta)$. **(b)** $(3 - \delta, 3) \cup (3, 3 + \delta)$. **(c)** $(-\delta, 0) \cup (0, \delta)$.

6. D_f contains $(p + 1, \rightarrow)$ and in general $(\delta, \rightarrow)$ for every real number $\delta > p$; each of these sets is a δ-set associated with $\underset{\rightarrow}{lim}$. Hence, D_f contains at least one δ-set.

10. D_f contains $\left[\left(\dfrac{a+b}{2}, b\right) \cup \left(b, \dfrac{b+c}{2}\right)\right]$ and in general, $[(b - \delta, b) \cup (b, b + \delta)]$ where $0 < \delta < \min\,[c - b, b - a]$; each of these sets is a δ-set associated with $\underset{b}{lim}$. Hence, D_f contains at least one δ-set.

12. For every real number $\delta > 0$ there exists a real number x such that $x > \delta$. Hence, $x \in (\delta, \rightarrow)$; thus Δ is non-empty.

16. For every $\delta > 0$ there exists a real number x $\left(\text{for example, } x = b + \dfrac{\delta}{2} \text{ or } b - \dfrac{\delta}{2}\right)$ such that $0 < |b - x| < \delta$. Hence, $x \in [(b - \delta, b) \cup (b, b + \delta)]$; thus, Δ is non-empty.

3-3 The Generalized Limit

2. Let any $\epsilon > 0$ be given. Since $|f(x) - c| = |c - c| = 0$, then $|f(x) - c| < \epsilon$ for $x \in \Delta$ where Δ is any δ-set. Since D_f is the set of real numbers, then D_f contains every δ-set. Thus, for every $\epsilon > 0$ there exists a δ-set Δ such that $|f(x) - c| < \epsilon$ if $x \in \Delta$.

4. Since *limit* $g(x) = -p$ and $-p < 0$, we can assert the following: For $\epsilon = \dfrac{p}{2}\left(\dfrac{p}{2} > 0\right)$ there exists a δ-set Δ such that $-p - \dfrac{p}{2} < g(x) < -p + \dfrac{p}{2}$. Thus, since $-p + \dfrac{p}{2} = -\dfrac{p}{2}$, there is a δ-set such that $g(x) < -\dfrac{p}{2} < 0$ if $x \in \Delta$.

6. *Limit* $f(x) = L$ iff for every $\epsilon > 0$ there exists a δ-set Δ such that $|f(x) - L| < \epsilon$ if $x \in \Delta$. But *limit* $(f(x) - L) = 0$ iff for every $\epsilon > 0$ there exists a δ-set Δ such that $|(f(x) - L) - 0| < \epsilon$ if $x \in \Delta$. Since $|f(x) - L| < \epsilon$ is equivalent to $|(f(x) - L) - 0| < \epsilon$, then *limit* $f(x) = L$ iff *limit* $(f(x) - L) = 0$.

3-4 Generalized Limit Theorems

2. We are given that *limit* $f(x) = L$ and *limit* $g(x) = M$. Hence, for any $\epsilon > 0$, there exist δ-sets Δ_1 and Δ_2 such that

$$|f(x) - L| < \sqrt{\epsilon} \quad \text{if} \quad x \in \Delta_1,$$

and

$$|g(x) - M| < \sqrt{\epsilon} \quad \text{if} \quad x \in \Delta_2.$$

Hence, $|(f(x) - L)(g(x) - M)| < \epsilon$ if $x \in \Delta_3$ where $\Delta_3 = \Delta_1 \cap \Delta_2$. Thus, by the generalized limit definition, $limit\ ((f(x) - L)(g(x) - M)) = 0$. To complete the alternate proof of Theorem 3.4d, we state:

$$0 = limit\ (f(x) - L)(g(x) - M)$$
$$= limit\ f(x)g(x) - LM.$$

Thus, $LM = limit\ f(x)g(x)$.

4. Let any $\epsilon > 0$ be given. We are given that there exists a real number $B > 0$ and a δ-set Δ_1 such that $|G(x)| < B$ if $x \in \Delta_1$. Since $limit\ F(x) = 0$ and $\dfrac{\epsilon}{B} > 0$, there exists a δ-set Δ_2 such that $|F(x) - 0| < \dfrac{\epsilon}{B}$ if $x \in \Delta_2$. Notice that

$|F(x)G(x) - 0| = |F(x)||G(x)|$ and $|F(x)||G(x)| < \dfrac{\epsilon}{B} \cdot B = \epsilon$ if $x \in \Delta_1 \cap \Delta_2$.

Choose $\Delta_3 = \Delta_1 \cap \Delta_2$ and we conclude that $|F(x)G(x) - 0| < \epsilon$ if $x \in \Delta_3$. Thus, by the limit definition, $limit\ F(x)G(x) = 0$.

3-5 More on Composition and Continuity

2. The functions whose equations are $y = x^2 - 1$ and $y = x^2 + 1$ are continuous at every real number b since they are polynomial functions. Since $x^2 + 1 > 0$, then by part (d) of Theorem 3.5b, $y = \dfrac{x^2 - 1}{x^2 + 1}$ is the equation of a function that is continuous for every real number b.

4. $\underset{2}{Lim}\ (x - 1) = 2 - 1 = 1$, and f is continuous at 1. Hence, $\underset{2}{lim}\ \sqrt{x - 1} = \sqrt{1} = 1$.

6. Note that $(2/x) \sin x = 2\dfrac{\sin x}{x}$ and $\underset{0}{lim}\dfrac{\sin x}{x} = 1$ (Example 3 of 2-15); hence, $\underset{0}{lim}\ ((2/x) \sin x) = \underset{0}{lim}\left(2 \cdot \dfrac{\sin x}{x}\right) = 2 \cdot 1 = 2$. Since f is continuous at 2, $\underset{0}{lim}\ \sqrt{(2/x) \sin x} = \sqrt{2}$.

8. The function $g = \{(x, x - b)\}$ is continuous at b, $g(b) = b - b = 0$, and the cosine function is continuous at 0 (Example 2 of 2-15). Hence, by Theorem 3.5d, G is continuous at b.

10. (a) $\underset{0+}{Lim}\ g(x) = \underset{0+}{lim}\ x = 0$ and $\underset{0-}{lim}\ g(x) = \underset{0-}{lim}\ 0 = 0$. Hence, $\underset{0+}{lim}\ g(x) = \underset{0-}{lim}\ g(x) = 0$, and $\underset{0}{lim}\ g(x) = 0$. (b) $\underset{0}{Lim}\ f(z) = \underset{0}{lim}\ 1 = 1$, but $f(0) = 2$; hence, $\underset{0}{lim}\ f(z) = 1 \neq f(0)$. (c) The composite function f of g is defined by

$$f(g(x)) = \begin{cases} 1 \text{ if } x \geq 0, \\ 2 \text{ if } x < 0. \end{cases}$$

Hence, $\underset{0+}{lim}\ f(g(x)) = \underset{0+}{lim}\ 1 = 1$ and $\underset{0-}{lim}\ f(g(x)) = \underset{0-}{lim}\ 2 = 2$. But since $\underset{0+}{lim}\ f(g(x)) \neq \underset{0-}{lim}\ f(g(x))$, then $\underset{0}{lim}\ f(g(x))$ does not exist; and f of g is not continuous at zero.

3-6 Limits in High School Mathematics

2. Assume $lim\ r^n$ exists when $|r| > 1$; that is, assume $lim\ r^n = L$ where L is a real number and $|r| > 1$. Since $|r| > 1$, then $0 < \left|\dfrac{1}{r}\right| < 1$ and hence, $lim\left(\dfrac{1}{r}\right)^n = 0$. Since $r^n\left(\dfrac{1}{r}\right)^n = 1$, then $lim\left(r^n\left(\dfrac{1}{r}\right)^n\right) = 1$; but by Theorem 3.5d $lim\left(r^n\left(\dfrac{1}{r}\right)^n\right) = L \cdot 0$. This implies that $1 = L \cdot 0$, which is impossible; hence, our assumption is false and $lim\ r^n$ does not exist.

4. Let any $\epsilon > 0$ be given. Since $lim\ s_n = L$, there exists $\delta > 0$ such that $|s_n - L| < \dfrac{\epsilon}{2}$ if $n > \delta$. Choose $\delta' = \delta + 1$; so that if $n > \delta'$, then $n - 1 > \delta$ and $n > \delta$. Therefore, $|s_n - L| < \dfrac{\epsilon}{2}$ and $|s_{n-1} - L| < \dfrac{\epsilon}{2}$ if $n > \delta'$. From properties of inequalities, it follows that $|s_n - s_{n-1}| < \epsilon$ if $n > \delta'$. But $|s_n - s_{n-1}| = a_n - 0$. Therefore, for every $\epsilon > 0$ there exists $\delta > 0$ (found as δ' above) such that $|a_n - 0| < \epsilon$ if $n > \delta$; $lim\ a_n = 0$.

6. Note that $\dfrac{\frac{4}{3}\pi(r + h)^3 - \frac{4}{3}\pi r^3}{h} = \frac{4}{3}(3\pi r^2 + 3\pi rh + h^2)$. Hence,

$$\lim_{0^+} \frac{\frac{4}{3}\pi(r + h)^3 - \frac{4}{3}\pi r^3}{h} = \lim_{0^+} \left[\tfrac{4}{3}(3\pi r^2 + 3\pi rh + h^2)\right] = \tfrac{4}{3}(3\pi r^2) = 4\pi r^2$$

because $\lim_{0^+} h = 0$ and π and r are constants for the given circle.

3-7 On From Here . . .

2. Let any $M > 0$ be given. Let $\delta = \sqrt{\dfrac{1}{M}}$. Hence, if $0 < x < \delta$, then $0 < x^2 < \dfrac{1}{M}$ and $\dfrac{1}{x^2} > M$; thus, g increases without bounds at 0.

4. The tangent to circle O at the point A.

Graphs of
Examples to Ponder

1–2 SOME EXAMPLES TO PONDER

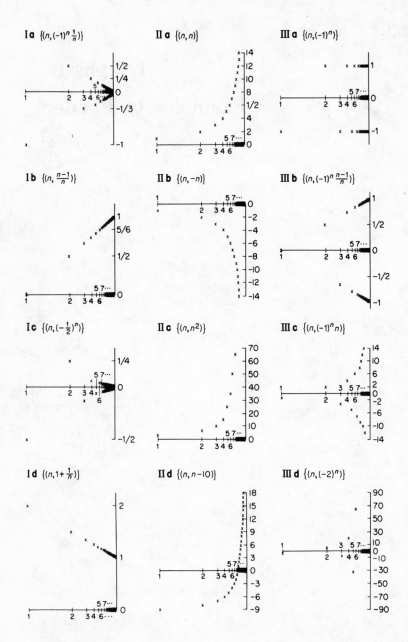

I a $\{(n, (-1)^n \frac{1}{n})\}$

II a $\{(n, n)\}$

III a $\{(n, (-1)^n)\}$

I b $\{(n, \frac{n-1}{n})\}$

II b $\{(n, -n)\}$

III b $\{(n, (-1)^n \frac{n-1}{n})\}$

I c $\{(n, (-\frac{1}{2})^n)\}$

II c $\{(n, n^2)\}$

III c $\{(n, (-1)^n n)\}$

I d $\{(n, 1 + \frac{1}{n})\}$

II d $\{(n, n - 10)\}$

III d $\{(n, (-2)^n)\}$

2–3 SOME EXAMPLES TO PONDER

Ia $\{(x, \frac{1}{x})\}$, $x \neq 0$

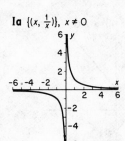

IIa $\{(x, \frac{1-x^2}{x})\}$, $x \neq 0$

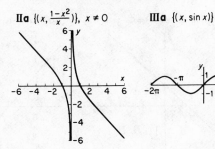

IIIa $\{(x, \sin x)\}$

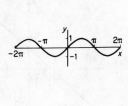

Ib $\{(x, \frac{x-1}{x})\}$, $x \neq 0$

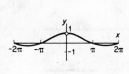

IIb $\{(x, \frac{x^2-1}{x})\}$, $x \neq 0$

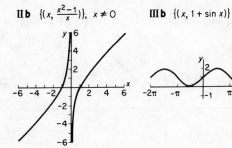

IIIb $\{(x, 1 + \sin x)\}$

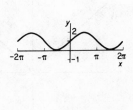

Ic $\{(x, \frac{\sin x}{x})\}$, $x \neq 0$

IIc $\{(x, x)\}$

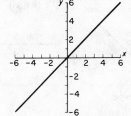

IIIc $\{(x, \cos x)\}$

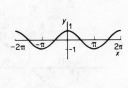

Id $\{(x, \frac{|x|}{x})\}$, $x \neq 0$

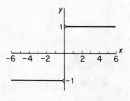

IId $\{(x, 2^x)\}$

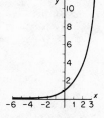

IIId $\{(x, |\sin x|)\}$

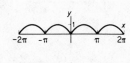

2–10 MORE EXAMPLES TO PONDER

Ia $\{(x, \frac{x^2+x}{x})\}$, $x \neq 0$

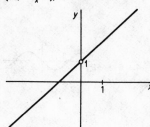

IIa $\{(x, \frac{|x|}{x}(x+1))\}$, $x \neq 0$

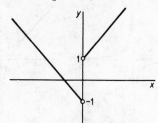

Ib $\{(x, |x|)\}$, $x \neq 0$

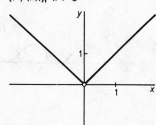

IIb $\{(x, \frac{|x|}{x}(x+\frac{1}{10^6}))\}$, $x \neq 0$

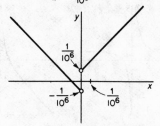

Ic $\{(x, y)\}$, $y = \begin{cases} x \text{ if } x \text{ is irrational} \\ 0 \text{ if } x \text{ is nonzero rational} \end{cases}$

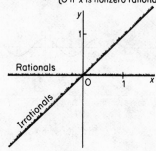

IIc $\{(x, y)\}$, $y = \begin{cases} 1 \text{ if } x \text{ is irrational} \\ 0 \text{ if } x \text{ is nonzero rational} \end{cases}$

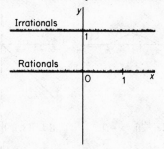

Index